PRAISE FOR
DRINK BEER, THINK BEER

"In his latest book, John Holl invites discerning drinkers to join him in a frank conversation about craft brewing's recent successes and future challenges. Brevity and pithiness are two of its biggest strengths, and the fact that it covers so much ground means you'll finish the last page with, as promised, plenty to think about. I'd pair it with an Old Ale, a style well suited for sipping and contemplation."

—Ben Keene, editorial director of *Beer Advocate*

DRINK BEER,
THINK BEER

JOHN HOLL

DRINK BEER, THINK BEER

GETTING *to* *the* BOTTOM *of* EVERY PINT

BASIC BOOKS
New York

Basic Books
Hachette Book Group
1290 Avenue of the Americas, New York, NY 10104
www.basicbooks.com

Printed in the United States of America

First Edition: September 2018

Published by Basic Books, an imprint of Perseus Books, LLC, a subsidiary of Hachette Book Group, Inc. The Basic Books name and logo is a trademark of Hachette Book Group.

Print book interior design by Amy Quinn.

The Library of Congress has cataloged the hardcover edition as follows:

Names: Holl, John, author.
Title: Drink beer, think beer : getting to the bottom of every pint / John Holl.
Description: New York : Basic Books, [2018] | Includes bibliographical references and index.
Identifiers: LCCN 2018008441 (print) | LCCN 2018012367 (ebook) | ISBN 9780465095537 (ebook) | ISBN 9780465095513 (hardcover)
Subjects: LCSH: Beer. | Drinking customs. | Brewing—Amateurs' manuals.
Classification: LCC TP577 (ebook) | LCC TP577 H65 2018 (print) | DDC 641.2/3—dc23
LC record available at https://lccn.loc.gov/2018008441
ISBNs: 978-0-465-09551-3 (hardcover), 978-0-465-09553-7 (ebook)

LSC-C

10 9 8 7 6 5 4 3 2 1

For Hannah. I'll explain all of this to you
on your 18th birthday at a pub in London.
We'll discuss it further at a brewery here in
the United States on your 21st.

CONTENTS

INTRODUCTION

"A bottle of beer contains more philosophy
than all the books in the world."
—Louis Pasteur

FOUR DECADES AGO, A FEW PIONEERS TOOK RISKS WITH BEER. Thanks to them, and to the consumers who wanted choice and thus supported their efforts, a brewing culture exists in America today that not only creates and supports local drinking communities but has launched a global phenomenon. More breweries currently operate in the United States than at any other time in our country's history. Barring an extinction-level event, the number should continue to rise for the foreseeable future. More breweries mean more beers, and more opportunities both to travel for a pint and to drink local. The industry's growth also means more experimentation with the world's second-most popular beverage (coffee has beer beat) and therefore more options—and more confusion—every time you step into a bar.

While it's a great time to be a beer drinker, the sheer volume of available choices can be overwhelming, even to the most experienced beer enthusiast. (Trust me.) For folks who are only moderately plugged into what's happening in the world of water, malt, hops, and yeast, it often seems easiest to default to familiar choices from large breweries that make their products approachable and relatable, thanks to heavy advertising and ubiquitous placement on shelves and taps.

That said, in the same way that many folks are rediscovering the importance of eating food produced locally, knowing where that food comes from, and getting adventurous when cooking at home, a similar principle holds true for beer. We can settle for the status quo, or we can branch out and experiment. When it comes to flavors in beer, everything is on the table: exotic fruits and vegetables, proteins, wood, herbs, and even a few things too gross to mention this early in the book (hint: yes, some brewers use animal organs and other animal parts in beer). A bit of time and a little education can open up a whole new world of beer for even the most casual of drinkers. Finding flavors that suit a mood, situation, or individual palate makes beer a uniquely personal adventure, just like discovering a favorite dish at a restaurant.

In no small part because of this experimentation, breweries have become destinations. You'd be hard pressed to find a general travel guidebook that doesn't mention at least one. Couples and friends build vacations around brewery visits, and enthusiasts will rise very early in the morning and line up outside a brewery to buy a limited batch of beer, as if it were an iPhone. Sierra Nevada Brewing Company's complex in Mills River, North Carolina, has earned the nickname Malt Disney World, because of the awe and childlike glee its gleaming facilities bring out in adults. Brewers are treated like rock stars. Fans queue up at festivals for the chance to have a beer poured by their heroes, for a selfie, for a fist bump.

I'm one of those fans (just maybe not the fist-bump part). I'm someone who enjoys a well-made pint, who gets lost in the appearance of an amber-colored IPA, watching bubbles soar with purpose from the bottom of a glass to a ceiling of foam. The kind of drinker who gets wide-eyed and happy with the first sniff of sweet, strong brown liquor rising from a barrel-aged imperial stout. I'll scratch my head trying to figure out the very specific flavor that comes and goes on the back of my taste buds—hot pepper, thyme, coffee—and will excitedly talk flavor and nuance with fellow enthusiasts until last call.

But I'm also a journalist. I started working in newsrooms at the age of sixteen as an intern for a local public television station that aired a nightly newscast. From there I moved on to newspapers, including eight years at the *New York Times,* where I spent a good chunk of my career covering crime and politics (often the same thing). Each day brought a new story, new people to interview, new cities to explore. That was what I most enjoyed about the job: I woke up each morning knowing I was going to work, but not what my assignment would be. (Now that I'm covering beer, getting up in the morning can be more difficult, depending on the night before.)

Three months before my twenty-first birthday my friend, Marc Cregan, gave me a subscription to a beer-of-the-month club. Every month six bottles arrived; I'd chill them down and try to drink them. Typically they were bottles from New Hampshire's Smuttynose Brewing Company, and because I respected this friend and his tastes, I was committed to giving each of them a shot. Dismayed, I inevitably admitted defeat in the presence of the hop bombs or other boozy concoctions and dumped them down the drain. But I was already intrigued about beer. I grew up in a house where my dad drank Heineken while the rest of the family downed Bud, or fondly reminisced about the days when Newark, New Jersey, breweries like Ballantine and Pabst were ascendant.

On the day I turned twenty-one, I walked into my local brewery. There was (and still is) a brewpub in my college town that had (and still has) an English pub feel, a popular theme with many breweries in the nineties. There, at the Gaslight Brewery in South Orange, I ordered a beer that went by three letters—IPA—and choked it down. The bartender—a man named Jeff Levine who would later become a friend—tried not to laugh as I forced myself to finish it. Against my better judgment I ordered another.

This time I asked about the beer, and Jeff gave me a brief lesson in hops. Pine. Grapefruit. Those are familiar aromas and flavors, yes? They are supposed to be in the beer. The bitterness is part of the experience, he told me.

I left feeling buoyed by my new knowledge, and ready to learn more. As I started traveling for the newspaper I worked for, I found myself searching out a local brewery wherever I landed. Often in those days it was a brewpub, and I'd wind up there without fail for dinner or a nightcap. I did this for three reasons:

1. Brewpubs had better food and beer than the lobby bar of whatever Holiday Inn I was staying at.
2. Local places gave me a more accurate feeling about the vibe of the town or city, adding important nuance to whatever human tragedy I was there to write about.
3. I could keep learning about beer.

I was purely learning as a fan back then. It took years before I felt comfortable writing as an "authority" on the subject. Soon enough I had graduated from covering crime and politics and transitioned to writing about beer. I started by working for small print operations like *Ale Street News* and *Celebrator Beer News,* monthly papers that covered the East and West Coasts respectively and were run by industry veterans and staffed by longtime freelancers. I learned about the nuance of local beer reporting, and the proper ways to write about beer flavors. I became a news editor at *Ale Street* and loved the alt-weekly vibe of the coverage: getting into odd corners of the industry, interviewing still-obscure brewers, trying to spot trends among the dozens of press releases that arrived daily. In 2009 I was hired as associate editor of the short-lived national publication *Beer Connoisseur Magazine.* While there I profiled beer-industry figures like the heads of Samuel Adams and Dogfish Head, and broadened my coverage and reviews to include international beer. In 2013 I was named editor of *All About Beer Magazine.* Founded in 1979, it's the country's oldest beer publication. I spent nearly five very happy and productive years reinventing the magazine's pages and collaborating with talented writers to craft in-depth and illuminating stories, all while working with some of the finest people I've ever had the privilege of calling colleagues. I also traveled the world, writing and reporting from five continents and nearly every state. In 2017 I was named senior editor of *Craft Beer and Brewing Magazine,* where I continue to cover the beer industry as well as home brewing, the hobby and passion that really kicked off today's beer renaissance.

Along the way I've hosted or cohosted several beer-themed podcasts and covered the industry for publications like the *Washington Post* and *Wine Enthusiast Magazine.* I've even written a few other books, including the *American Craft Beer Cookbook,* which celebrates pairing good beer with good food.

These days, as a journalist who covers beer full-time, I have the privilege of often being front and center for truly remarkable events and conversations throughout the brewing industry. My access—along with a healthy curiosity, exhaustive travel, and research trips around the world and around the corner, coupled with my duty to share what I've learned from professionals, fellow drinkers, and quite a few history books—has led to the book in your hands.

Beer is more than a combination of ingredients in a pint glass. There's a whole orbit of other elements around that glass to acknowledge, appreciate, and understand. Beer has an economic impact. It's a social convention. There are family ties to beer, fond (or not) memories of what your parents and grandparents drank. Conversations with colleagues about humorous (or not) commercials that the larger breweries have plunked into the Super Bowl. A nonstop blur of billboards along our highways, neon signs in tavern windows, jingles on the radio. And more often than not, that one friend or cousin who talks only about beer when you get together.

Beer is the story of progress and hard work. It's farmers working the land for ingredients. It's activists fighting to keep water sources pure. Though you may not have learned about beer in your history class, it plays a bit part, though a vital one, in world history, from the Pilgrims, who arrived on the *Mayflower* with beer as their potable water, to the Industrial Revolution, when technological advancements in refrigeration made it easier to transport and store beer. Beer has even seeped into politics and university classes. It's no longer confined to its own industry.

Thanks to a strong American influence, a global beer renaissance is in process. Countries with long, proud brewing traditions are following the Yankee lead and innovating on classic styles and recipes.

In America and abroad, it's impossible to escape beer in popular culture. Tom T. Hall twanged his way through a song about how it helps him unwind and feel mellow. Red Solo cups are the vessel of choice for kegged beer at a weekend cookout. Laverne and Shirley remain the pop-culture patron saints of everyone who works along a bottling line, and once-fictional brands like Duff can now be consumed by visitors to the Simpsons area at the Universal theme parks. Oh yeah!

In truth, although it is a great time to be a beer drinker, it's also a confusing time. There are poorly made beers, misinformation about flavors, and perhaps too much choice. For every public relations company and industry association smiling and shooting sunshine, there's a dark side that involves pay-for-play with accounts, access to ingredients denied, unsafe working conditions, and undercurrents of racism and sexism in an industry that seems to still favor white males above all others.

To drink beer is easy. Pour, put to mouth, swallow. To think about beer is much harder. It's a social beverage that has long brought people together, from friends gathering after work to blow off steam, to revolutionaries planning war. For a growing number of others, it's an integral part of life that extends to online chat rooms, vacations revolving around breweries, and a never-ending quest to taste only the rare beers—the "whales"—that exist in the smallest of quantities. Although fun for some (and certainly profitable for certain breweries), this kind of beer-drinking-as-sport hasn't done the larger industry any favors. A few drinkers of wine or spirits will turn their noses up at the very thought of beer, refusing to take the beverage seriously. Honestly, who can blame them?

The brewing industry just before and certainly after Prohibition gave us mainly one style of beer. Think of it as beer-flavored beer—generic, easy drinking, and generally lacking in discernible flavor. For all it was to some, it was less to others. Beer, inexpensive and ubiquitous, was considered low-class or even trashy.

Following Prohibition, wine asserted itself at the fine-dining table. And spirits became associated with high-end good times via cocktails. People who still look down on beer despite how far it has come since the 1970s may do so because of a long-stale social stigma, a bad drinking experience in college, the overwhelming bitterness of some styles, or maybe they're just buying into the marketing against it. If you know someone who falls into this category, it's time to start changing their mind: the days of beer-flavored beer are behind us.

Despite being around for several millennia, beer finally seems to be coming into its own. Science helps brewers make better beer. Technology assists farmers to grow barley and grains more efficiently and to develop new varieties of hops. Laboratories learn more about yeast with each passing month, capturing and cultivating different strains to add new flavors to beer. Some labs are working to grow strains of wild yeast that have been plucked from specific zip codes, meaning that you can literally drink a local beer.

Now, after a rocket-like ride to this moment in time, it's hard to ignore the endless waves of beer—both good and bad—that shape the everyday drinking experience for tried and true fans, as well as for folks slowly coming around to the taste of beer. Nearly seven thousand breweries currently operate in America. A few are probably located in your neighborhood. More are on the way.

Beer doesn't come in only one flavor, and we've certainly moved beyond the options that simply "taste like beer." Thanks to the seemingly endless supply of flavors, I firmly believe that beer pairs better with every type of food than any other alcohol. Yes, even wine. Restaurants seem to agree; beer in 750-ml bottles or on draft has found its way onto fine-dining menus countrywide, and high-end beer is replacing or joining the long-standing generic lagers at taco trucks, burger joints, and greasy spoons, creating a more rounded culinary experience.

Just as modern technology has changed how we make beer, social media has launched us into a global pub that never closes, and where everyone, it seems, has an equal voice. Entire websites and virtual communities exist to talk about, review, dissect, gossip about, trade, and obsess over beer. From the sometimes snarky and opinionated Beer Advocate (which spawned a magazine and multiple festivals) to the more analytical and uppity RateBeer (now partly owned by the venture-capital arm of Anheuser-Busch InBev) to the ubiquitous Untappd (a mobile-phone app that allows drinkers to "check in" to a beer), finding like-minded barstool philosophers, unsolicited advice, or sympathetic gadflies is as close as your fingertips.

Beer isn't binary. It's continuously evolving, and as such is the topic of countless debates over what beer is and isn't, and what it can be. Once it was an accepted fact that a beer with a skunky aroma (you've had a Heineken at some point, yes?) was flawed, the victim of sunlight negatively impacting hops through the glass. But now some brewers are pushing back; they think (correctly) that hints of a lightstruck aroma might actually help improve certain beer styles, like a saison.

All the new breweries that opened and all the new flavors released to the marketplace needed a way to stand apart from the pack, to establish themselves as separate from the historical norms. Thus, for the better part of two decades, the word *craft* was attached to beer to imply some kind of gravitas: a higher quality than mass-produced beverages from large corporations. Now those giant companies, like Anheuser-Busch InBev, are buying smaller breweries and adding them to already robust portfolios. This phenomenon has emboldened the makers of Budweiser (and other brews) to use the word *craft* liberally. As a result, the smaller brands have gone in search of a new moniker.

But at the end of the day it's all still beer. It is what people gather over at pubs. What they use to toast both success and loss. To celebrate victories, or to imbibe as a means of escape. It's what we curse in the morning after too many pints. It's so much more than just a liquid. As a friend correctly pointed out to me one night after several rounds, beer is an addendum to life. And the more we acknowledge, appreciate, and understand about beer, its history, and its social place in our world, the more fulfilling a drinking experience we'll have.

In the four decades since the first post-Prohibition brewery opened, the industry learned to run before it knew how to crawl, and now it's going so fast that few stop to think about the ramifications of such a quick evolution. Progress for the sake of progress is rarely good, and beer is in danger of losing its way, its soul, its importance, and its identity as separate from other alcoholic beverages. Overcommercialization, a push to accommodate humans' natural inclination for sweeter tastes, and a deliberate willingness to chase the next shiny thing have led some brewers to forge uncharted paths, guiding drinkers, typically the younger ones, away from beer's roots. In response, there's another faction of beer drinkers—we can call them purists, but they are basically seasoned drinkers who have been at this for a while—who shout at the youngsters that the new beer styles they are drinking are essentially the emperor without clothes.

I believe that beer can be romantic, that the relationship between glass and drinker is special—from the mere smell of a well-made beer popping with pleasant aromas, to the feel of a snifter of barley-wine in your hand while cozied up at a bar or ensconced in your favorite chair with a good book on a winter's night, to the satisfying finish that leaves you full and perhaps a little contentedly intoxicated. The sound of a corked beer being opened is a happy one, as is the clank of glasses among friends toasting a special occasion.

For everything that's special about beer, we're at the point where it becomes something less. People go to Starbucks or Dunkin' Donuts out of habit and convenience, not necessarily because they have a deep love of the product. It's just what you do, and the coffee or pastries those companies turn out perform the job well enough. Certain beers fit that mold—for instance, Bud Light—but just because they exist and are popular, is that where our attention should be? I don't think so. So many high quality beers are being made (and so many to suit the tastes of every drinker) that there's no longer a need to settle. You can spend an entire year drinking a different IPA each day without scratching the surface of all the IPAs that are out there. Or you can drink the same beer regularly, like a Belgian-style tripel, and probably discover something subtly distinctive each time. Drinking beer doesn't need to be top of mind each time you take a sip, but each beer should be enjoyable, not just for your taste buds but also for your mood.

So what is this book? Unlike many introductions to the topic, it is not solely focused on the sensory experience of drinking beer. I believe that the sensory element is important, that everyone who eats or drinks anything should consider the processes, the flavors, the sensations with each bite and sip. Doing so leads to a fuller involvement, especially if we're in the habit of eating and drinking merely for fuel rather than for pleasure. But over the last sixteen years of writing about beer, what I've become obsessed with is the peripheral: the experience not only of the beer itself, but of all that goes into its creation, and where it takes us both personally and together.

This book aims to talk about the fact that having a beer isn't quite as simple as just having a beer. Dozens of factors influence your beer decision before you even order and take that first taste. In the pages that follow I'm excited to pull back the curtain on those things: from how the first modern beer pioneers shaped the flavors of the beers we drink today, to the way advertising continues to play a pivotal role in sales.

I've walked the aisles of beer shops and grocery stores looking at labels and logos, and I've seen how the right color, design, and placement move certain beers. I've sat at countless bars and watched the aesthetics of drinking—from logo design to the shape of tap handles to different glassware—work its magic. I've attended parties where I've observed social pressures placed on someone to drink a beverage that will convey a certain message. We all experience these events, but we rarely stop to think about them.

To fully understand and appreciate beer, we also need to consider the raw ingredients—not just as they arrive at the brewery or appear in a final recipe, but how they are grown, farmed, processed, and ultimately taste, both on their own and in terms of how they impact other flavors. Even a modest understanding of ingredients leads to a better drinking experience. The same is true for knowing the basics of beer service, health impacts, and stigmas. Being a well-informed beer drinker doesn't mean being a snob. I've spent a career drinking excellent beers (some dreadful ones, too) and learning from the pros. Firsthand I've witnessed certain childish and troubling aspects of the industry—the way some breweries treat women or minorities, or how they knowingly serve substandard beers. But I've also seen the very best of humanity and kindness thanks to this remarkable beverage.

The brewing industry is on full display daily at the thousands of breweries and brewpubs across America. There, you can witness not only the science and the process of beer-making and serving but also your neighbors and like-minded beer-drinking folks from out of town. Each pint ordered at one of the newer-generation breweries helps a small business and its local community.

Ideas are discussed and formed over beer, new passions unearthed, and new friends met. If you include beer as a part of your travel plans, you're likely to make discoveries that otherwise never would have appeared on your radar. Trust me, this is the experience talking. I've walked hop fields during harvest, and cleaned kegs at my local brewery. I've homebrewed with friends and helped on batches with the pros. I've logged thousands of hours on barstools and standing among stainless steel fermenters with a notebook in hand—all because I love my job and want to understand the subject better.

SHERLOCK, THE RECENT MASTERPIECE TELEVISION SERIES, INTRODUCED many to the concept of a "mind palace": a place where the famed detective would go to retrieve all manner of information that had been locked away in his brain. As you progress through these pages, I'm occasionally going to ask you to visit the "mind pub." It's the bar that exists only in your head and should be the perfect (to you) representation of all you want from a drinking establishment. Everything from the decor to the music to the location is up to you. It doesn't even have to be an actual pub. All I ask is that there are at least a few taps on the wall and some glassware available.

Beer, like life, is ever evolving, and each day brings new findings, new flavors, new people. What I've learned and hope to communicate is that beer is more than just the contents of a glass. I believe that the greater our understanding and appreciation for what is happening *outside* the glass, the better the impact on what's in it.

Let's settle into a pint and discuss what makes beer so great.

ONE

THE MODERN BEER RENAISSANCE

EVERY TIME YOU BUY A PINT OR PURCHASE A SIX-PACK YOU'RE choosing a side in a war.

In this "us versus them" conflict, the "us" is typically smaller breweries, who proclaim themselves dedicated to the craft of beer-making. "Them" are the global and massive breweries like Anheuser-Busch InBev, Heineken, Miller, or Molson Coors, who, so the story goes, care more about profits than consumers or taste.

But even within Team Small Brewery there is constant competition for the finite amount of tap-handle space. The winners are chosen by consumer dollars, and the tactics can be extremely dirty.

In many ways, modern brewing is shaped by this big guy/little guy conflict. Understanding how we got to this point requires a look back at the history of brewing in America. Although this country was founded on beer, we've had a complicated relationship with it from day one.

When I say America was founded on beer, that's not as much of a hyperbole as you might think. William Bradford, a settler who sailed over on the *Mayflower,* wrote in his diary in 1622 that the decision to land at Plymouth, despite an original plan to continue farther south, was made due to a supply problem.

"We could not now take time for further search or consideration, our victuals being much spent, especially our beer," he wrote. And so the hearty souls disembarked and got down to the business of brewing . . . and forging a new settlement.

This is nothing new for beer. Historians tell us that as far back as 15,000 BCE, in China and the Middle East, nomadic tribes began to settle only when the wild grasses they grew became permanent crops, allowing them a stable place to grow, harvest, and live. Although the results were quite different from the beers we'd know today, when those newly cultivated grains were mixed with water and then attracted the natural yeast in the air, the result was a grain-based alcoholic beverage. Many of the original brewers believed that the ensuing intoxication was the work of the gods, specifically in Sumeria, where the brewing goddess, Ninkasi, was routinely and rightly praised. Access to the drink (along with a geography that regularly provided food) helped form some of the world's first settlements.

From these humble origins we can quickly jump ahead through time, through the rise and fall of civilizations, past where ingredients like hops were discovered and became part of beer, where recipes were developed, where brewing became a profession and beer a national identity, and finally arrive back in the United States. During the American Revolution, the founders of the country regularly met in taverns to drink cider and ale as they planned to rebel against the monarchy and create the United States as we know it today.

Many of those revolutionaries are often cited as brewers. George

Washington's porter recipe is kept in the New York Public Library. Beer was regularly made at Monticello, Thomas Jefferson's home. In truth, though, as with so many household chores, it was the women—or the slaves—of the house who were responsible for making beer. As much fun as it'd be to think of the Founding Fathers standing around the garage on a weekend, sleeves rolled up, brew kettle going strong, fife music chirping in the background, it's simply not the case. That said, by both brewing beer and drinking it, the Founding Fathers and other revolutionaries sewed the importance of beer into our country's fabric.

A quick diversion: that quote you hear attributed to Benjamin Franklin, the one plastered on countless T-shirts, pint glasses, and posters hanging in breweries? "Beer is proof that God loves us and wants us to be happy"? He didn't say it. At least not like that. The written quote was originally about rain, how it nourishes vines and gives us *wine,* not beer. But despite this myth regularly being debunked by historians (along with so many of his other quotes), and me writing it here, I don't expect it to go away anytime soon. Plus, who among us hasn't been misquoted after a few rounds?

As America became a country of immigrants, many of those who came through Ellis Island sought a new life using their skills from the homeland, naturally. Farmers took to the fields, carpenters built, bakers baked. And brewers, of course, brewed.

Wherever people lived there were breweries, especially in river cities—Brooklyn, Paterson, Newark, Philadelphia, Boston, Chicago, St. Louis. The majority of these breweries were local. Transportation being what it was, citizens weren't able to travel far for what they needed. So if you lived within, say, ten blocks of a brewery, that would be your brewery. If you were eleven blocks away, there'd likely be one closer, and you'd frequent that one instead.

As on so many fronts, America has always had a complicated relationship with alcohol. When Prohibition hit in 1920, the anti-alcohol and pro-Puritan sentiment had been building for decades. We know, of course, that booze didn't fully disappear. And from its being pushed underground, we are left with enduring images of the Fitzgerald speakeasy era. However, the alcohol industry as a whole was largely decimated, especially hop crops and cider.

Roughly forty-eight hundred breweries were operating in the country when Prohibition began. Thirteen years later, when it was repealed through the passage of the Twenty-First Amendment, that number was down to only a few hundred. From there things got worse for consumers. Big brewers became bigger, either absorbing competitors or eliminating them outright through sales or better recognition. By the 1970s the number had bottomed out at fifty or so.

So how did we get to our current beer renaissance? Have you heard the theory that a butterfly flapping its wings on one side of the planet can create the wind that will become a hurricane on the other? In the case of modern beer in America, Jack McAuliffe is the butterfly.

If you were to draw a line from the craft-beer industry in America today back to its beginning, that line, despite any sharp turns, dips, and rises, would lead directly to McAuliffe. In 1976 he founded the New Albion Brewing Company and in a relatively short period changed the course of brewing in this country, and therefore worldwide, reviving a once-proud tradition that had sadly fallen into obscurity.

The late 1970s were an interesting time for beer. Homebrewing had become legal again, thanks to a law signed by Jimmy Carter in 1978 that removed the tax on beer made at home for personal use. It was one of the last provisions on the federal level left over from

Prohibition. In garages, basements, and backyards across America, creative souls fired up burners and began mixing ingredients from recipes that had been mostly abandoned by large brewing companies, which had chosen to focus on mass-produced lagers.

These homebrewers tried their hands at saison, porter, and barleywine, bringing long-lost flavors back to enthusiasts and those looking for something more than the generic, one-size-fits-all lagers that dominated American shelves and taps. Change, however, was starting to creep in, both domestically and internationally. Imported lagers—which stood out thanks to their unusual green bottles in a sea of brown—and other specialty beers brewed overseas were beginning to make shy appearances in American stores.

It was in Europe, where professional brewing had operated uninterrupted for hundreds of years, that McAuliffe learned both to appreciate good beer and to brew it. A US naval engineer stationed in Scotland and working on submarines, he frequented the local pubs, trying all the ales—especially stouts and porters—on offer. Interested in expanding his horizons beyond just drinking, he applied a practical method to his hobby, devouring every brewing book he could get his hands on. He eventually walked into a Boots pharmacy and purchased a ready-to-go homebrew kit. McAuliffe took it back to his rented house, mixed up the packaged ingredients in water, and let the blend sit and ferment. When it was ready to drink, he shared it with his fellow sailors, who gave him notes for his next batch. It was this combination of communal gathering and ingenuity that appealed to him.

When McAuliffe left the Navy and returned to the United States, he headed west and landed in Sonoma, California, where he picked up work building houses and laid plans for opening a brewery. He consulted textbooks and magazines at the University of

California at Davis, the country's premier brewing school, and put together his own beers—a pale ale, a porter, and a stout—based on historical recipes.

With just a few thousand dollars, and alongside his girlfriend and business partner, Suzy Stern, he opened the New Albion Brewing Company in 1976 inside a rented shack of a warehouse in Sonoma. Unable to find manufacturers that made smaller brew kettles, fermenters, and other equipment that would fit his company's modest size, McAuliffe, who had spent half his high school years in a welding shop, built his brewery utilizing stainless steel equipment that had once been part of dairy farms and soda-manufacturing businesses and that he picked up on the cheap . . . or just picked up.

He also manufactured his brewery's history. There had been an Albion Brewing Company operating in San Francisco before Prohibition, and he laid claim to their legacy with a label created by a local designer. It showed Sir Francis Drake's ship the *Golden Hind* sailing through the mountainous inlet into San Francisco Bay against a baby-blue background with sweeping script.

"History is important in the brewing industry," McAuliffe told me in 2010. "But if you don't have a history you can just make one up. We made English style ale, porter, and stout. The New Albion Brewing Company, get it? Name, logo, and history, bang!"

A BARREL OF BEER CONTAINS 31.5 GALLONS. MCAULIFFE'S SYSTEM could make one and a half barrels at a time. Compare that with, say, Schlitz, in Milwaukee, which during that same period was making nearly nine million barrels of beer per year, according to brewery statistics. That wasn't uncommon. The majority of the breweries operating at the time (save for a few, like San Francisco's Anchor Brewing

Company) were large, hulking factories turning out millions of barrels of beer—and were unlikely to make it onto anyone's vacation itinerary.

Enthusiasts who heard about New Albion through word of mouth or in local news articles regularly drove to the brewery to try this unique, small-batch beer that was made with just water (trucked in because the site didn't have a well), hops, malt (processed at the Anchor brewery in San Francisco because suppliers didn't deal with businesses the size of McAuliffe's), and yeast. He declined to use rice, an ingredient routinely used by large breweries, especially for lager. Interest in this upstart grew quickly. Visitors started showing up at his door, and bars in San Francisco placed orders for bottles. Before long the national press took notice, with reporters from both the *Washington Post* and the *New York Times* traveling to California to write up this new venture in brewing.

Those articles helped inspire others around the country to take up arms against the larger brewers and mark the return of localized brewing to the United States. Three people in particular from those early days helped change the course of American brewing.

The first was Dr. Michael Lewis, a professor at UC Davis who had helped McAuliffe with research as he sought to launch the brewery. Once it opened, Lewis drove there to see what McAuliffe had created.

Prior to 1977 the brewing program at the large California university had existed to train students—mostly men—how to brew on a very large scale. There were only fifty or so breweries operating in the United States at the time, the majority of which were the behemoths owned by companies like Coors, Miller, and of course Anheuser-Busch. The beers they produced, with only a few exceptions, were in the American lager style: essentially that beer-flavored beer. The graduates of the program at UC Davis were more likely to be working in white lab coats with sophisticated equipment suitable for large-scale manufacturing and precision science than they were expressing themselves artistically. When Lewis saw what McAuliffe had fabricated, as well as the wild consumer response to his product, he worked to change the school curriculum to accommodate those who wanted to work at smaller breweries or to start their own. Today dozens of brewing schools exist around the country, along with countless brewing programs at colleges and universities. The graduates of these programs continue to fuel extraordinary growth in the industry. In fact, current graduates of the UC Davis program are more likely to work at a small brewery than a large one. Lewis told me that McAuliffe "certainly changed my view of what the industry could be. I saw a new direction for the industry and a new direction for my program. Jack was the beginning of that."

The second important person to show up at McAuliffe's door was Don Barkley, a young man who had started homebrewing in high school and was thinking of a career in brewing. He showed up one morning in the office of New Albion and stated his intentions to work there for the summer. For free. McAuliffe—whom I'll politely describe as gruff in this situation (and others)—told Barkley to leave.

Undeterred, Barkley persisted. He returned later to find Stern. By this point the pair had been working eighty-hour weeks with few breaks. Free labor would be just fine, she said, overruling Jack. Barkley was paid with a case of beer per week, along with bottles on the job as desired. He pitched a tent on the property and worked long hours in cramped conditions. Officially known as the assistant brewer, in reality he did whatever was asked of him—from washing kegs and bottles to sweeping floors. And he did it cheerfully, grateful for the experience.

Barkley is the true patron saint of assistant brewers in this country. Assistant brewers are there to help head brewers or brewmasters; often their role involves performing grunt work on the brew deck, but it also allows for hands-on experience, the chance to create recipes and put to practice all that was learned in school. You're hard pressed to find a brewery in the country that doesn't have at least one assistant brewer, and Barkley is a man to thank for that. He's the inspiration for so many young people looking to forge a career in the industry by grabbing their bootstraps and making it happen. Barkley spent the majority of the rest of his career working at Mendocino Brewing, in Napa Valley. He is still involved with various brewing projects, and as a mentor, training and teaching future generations of brewers.

WE'RE NEARING SEVEN THOUSAND BREWERIES IN THE COUNTRY now. In my many conversations with brewers, I've often asked about the first beer that inspired them, that introduced them to flavor and the promise of something more than beer-flavored beer. Again and again, they've pointed to Sierra Nevada Brewing Company's pale ale. The soft blend of caramel malt and two-row malt with Perle, Magnum, and Cascade hops is an American classic, originally (and still) served in a heritage 12-ounce bottle with a distinctive green label that boasts "Purest Ingredients, Finest Quality." Sierra Nevada Brewing Company was cofounded in 1980 by Ken Grossman, the third important person who walked through New Albion's doors in those early days. Grossman's beers have inspired countless brewers, but he himself was in part inspired by McAuliffe.

Grossman discovered homebrewing when he was a teenager, thanks to a neighbor, and had always been fascinated with engineering and mechanics. He traveled from Chico, a town in northern California, to tour New Albion in Sonoma, and while examining the three-level, gravity-fed brewhouse that McAuliffe had constructed, he realized that he had seen similar equipment close to home and that he too could create a brewery.

Later, Grossman would say he didn't remember much else about the tour (McAuliffe said he didn't remember Ken at all), except for the noticeable hop content in the beer. Nonetheless, afterward he pulled together about $50,000, founded his own brewery, and began selling beer around Chico. Focusing on a commitment to quality control and an innovation on styles (the mission statement on the label), Sierra Nevada Brewing Company grew. Today it's the third-largest craft brewery in the country, behind D. G. Yuengling and Son and the Boston Beer Company, makers of Samuel Adams. In 2017 Sierra Nevada made 1.4 million barrels of beer between its original location and one it constructed in 2015 in Mills River, North Carolina.

In comparison, in 1980, New Albion's best production year, it topped out at 450 barrels. McAuliffe was never able to grow the brewery enough for it to become financially viable. After just five years he shuttered the company's doors. His equipment, along with Don Barkley, headed north to the Mendocino Brewing Company. Other parts of the business were scattered to the wind. The original brewery sign now hangs above the bar, in a spot of honor, in the taproom of the Russian River Brewing Company, in Santa Rosa, California. McAuliffe visited several years ago and signed it.

McAuliffe left brewing and was largely forgotten as the industry that he inspired took off without him. He worked various engineering jobs and occasionally made moonshine, but stayed out of the beer industry and the spotlight. Ken Grossman, however, always remembered. When it came time to celebrate his brewery's thirtieth anniversary, he reached out to McAuliffe, who was living in Austin, Texas, and invited him to brew a special beer. Grossman even flew to Texas to ask him in person. In a scene not unlike McAuliffe's first encounter with Barkley, McAuliffe told the younger brewer to leave. Undeterred, Grossman eventually convinced the pioneer to return to the brewhouse.

In May 2010 McAuliffe professionally brewed again for the first time since closing New Albion. He arrived on the morning of the brew day wearing jeans and a short-sleeved blue oxford shirt. As he shared his story with a video crew hired by Sierra Nevada to document the events, McAuliffe sipped on a Kellerweis, an American-style hefeweizen, often nodding in approval after each sip.

The actual brewing was a largely ceremonial affair. After a morning of interviews, McAuliffe and Grossman walked into the brewhouse, each grasping the handle of a green plastic pail filled with Brewers Gold hops, and tipped the cones into a large copper brew kettle. Sierra Nevada's capable staff handled most of the rest of the process. The two sampled the wort, which McAuliffe declared "very good," and then headed into the brewpub for lunch.

That day in 2010 was something of a turnaround for McAuliffe. Since his brewery had shuttered he had chosen to stay out of the spotlight, and he wasn't shy about sharing his frustrations that others had succeeded where he had not. In 2009 he had been involved in a serious car accident that severely injured his left arm, triggering a difficult stretch in his personal life. A few weeks after the brew day at Sierra Nevada, I was surprised to receive a phone call from Jim Koch, the founder of Boston Beer Company, makers of Samuel Adams.

Boston Beer Company is the second-largest craft brewery in America, and Koch has spent a career growing it and its subsidiaries (Angry Orchard Cider, Twisted Tea, and other beverages and smaller brewery brands) into a powerhouse. Although occasionally—and rightly—criticized by other brewers for what they've seen as Boston Beer's underhanded tactics (like stacking the consumer's preference poll at early iterations of the Great American Beer Festival), Koch has also respected the brotherhood of brewing and routinely stepped up to help others in need. During a hop shortage a decade ago, for instance, he supplied smaller brewers with his brewery's stock, and through programs like Brewing the American Dream, he helps small brewers and entrepreneurs secure micro loans for their businesses. The brewery also runs the annual LongShot homebrewing competition, the winners of which see their small batch brewed commercially. Several winners have since gone pro.

In his phone call to me, Koch told me that in 1993 he had taken ownership of the trademark for New Albion when it was about to expire, in order to protect the integrity of the first craft beer and preserve the name of the brewery. Koch reached out to McAuliffe and proposed reviving New Albion for a modern age. After months of negotiations and conversations, McAuliffe arrived at the Boston brewery on July 3, 2012, to mash in the first batch of New Albion Ale in thirty years.

They created a pale ale as "faithful" as McAuliffe could recall to his original. Malt variations that replaced varieties used in the 1970s were substituted, for example. More than six thousand barrels of the ale were made and released with the original label art. Once it sold out, it was gone. Koch gave all the profits from the sale of that beer to McAuliffe, providing him both a proper retirement and the payday from beer that he so richly deserved. The New Albion trademark also went back to the McAuliffe family. These days McAuliffe's daughter is contracting New Albion ale through Platform Beer Company in Ohio.

Years ago, Sam Calagione, founder of Dogfish Head Craft-Brewed Ales in Delaware, called the craft-beer industry "99 percent asshole-free." The number has likely dipped into the high 70s since then, as sales and beer popularity have soared, but the sentiment remains true. As much as there is increased competition among the thousands of breweries fighting for shelf space, tap handles, consumer dollars, and relevance, there's still something ingrained in brewing professionals: the desire to create, to be part of a community, and to see the profession grow. Sam Adams, and Jim Koch in particular, gets grief these days from smaller breweries and some drinkers for being more corporate than craft, but stories like their partnership with McAuliffe should remind folks that there are humans behind the business.

The breweries that have succeeded have done so largely because of people like these. Consumers like to identify a product with a person, someone they can relate to outside of the item itself. The country's larger craft breweries have relied on the reputations of folks like Koch, or Grossman, or Kim Jordan of New Belgium Brewing. The same is true all the way down, to the brand-new local brewery in your hometown. If you can walk in, meet the owners, meet the brewer, shake their hand, and feel good about that connection and who your dollars are benefiting, you'll care about the product a little more. And it goes both ways. Successful breweries are backed by people who actually care about their business and their beer, and who show it every day with each interaction.

SINCE THE DAY NEW ALBION OPENED TO THE PUBLIC, THE NUMBER of breweries in this country has steadily climbed. But not until 2016 did US breweries officially surpass, finally, pre-Prohibition numbers, according to a Brewers Association tally. Early in this renaissance the majority of breweries produced relatively small volumes and served modest geographical areas. The players that were big at the beginning of the period are even bigger now, especially Anheuser-Busch InBev, a company based in Belgium that through mergers and acquisitions has become the largest brewing company on the planet.

The big brewing companies didn't seem to pay much attention to the newcomers in the late 1970s and early 1980s. Business was good, so why should they? The majority of beer drinkers in the country preferred (and still do, actually) to purchase light, mass-produced lagers, so the smaller breweries didn't upset the Clydesdale-driven cart.

But change was coming. In the late 1990s and early 2000s, these "little upstarts" began to grow. And as they grew, they kept poking the bear—or, more accurately, they fired the first shots in the war. Koch of Boston Beer is famous for saying that the larger brewers spilled more beer at their breweries each year than he was able to produce at his. Through pointed advertisements he routinely questioned their methods and commitment, sparking the ire of the established players.

Across the country more and more new players tried to capitalize on the microbrewing industry. As with any emerging business, some folks got into it for the wrong reasons or were ill-trained to uphold industry standards. Worse still, gimmicky brands started popping up: breweries that went for a cheap laugh through silly beer names or eyebrow-raising labels designed to stand out on shelves and convince drinkers to pay a premium for a six-pack. Often the promotional strategies utilized a species of animal and a bodily fluid (Mouse Tears, anyone?), or a cheeky reference to a busty woman. Unfortunately, these attention-seeking labels were usually the best part of the beer. Because the creators focused primarily on marketing, the beverage suffered. Even in styles where it's easy to hide flaws and off flavors—like stouts—if the defects were too pronounced, and often they were, the liquid would soon find its way down the drain. (One notable exception from this era that still thrives today is Moose Drool Brown Ale from Montana's Big Sky Brewing Company, which is actually good.)

To give customers a chance to try the different beers on offer, the tasting flight was introduced. You could walk into a brewpub—often designed like a British tavern—and order a sampler tray of the house-made beers that showcased a diversity of color, clarity, and flavor. There'd be a blonde ale, an amber, a brown ale, a pale ale, and a stout. Perhaps a hefeweizen or barleywine. Flights remain commonplace today, although some breweries only offer one or two samples at a time rather than a whole complement. In the early days they were a novelty—a fun, visually interesting way to kill an hour and explore what a brewery had to offer. But the thing was, at some brewries almost all the beers in the flight tasted the same. Sure, the stout would (hopefully) have some chocolate or coffee flavors to it. There might be some discernible hop characterizing the pale. Maybe the hefeweizen would be cloudy. But in the background of taste, all the beers had a sameness quality.

The explanation is simple: some of the brewpubs had one or two types of yeast they used for fermentation—the house strain—and they added it to all their beers. That meant consumers were drinking mostly with their eyes, missing out on the nuances of flavor that come with the use of appropriate yeast strains. Think of it as a bakery that only made sugar cookies—then added different colors of frosting to different batches for visual variety.

Still, the beers on offer were something different, something that bucked the norm that had settled over American beer for generations. Soon enough the larger breweries noticed and started taking swipes at their very small competitors. Why would you want to drink *that*, they asked their massive and loyal customer base. You know us, they said. You've trusted us, your grandfather drank this brand. Why concern yourself with those upstart brews, with their odd flavors, their darker color—it's too heavy if it's that dark anyway—and the names you've never heard of? In 1996 a widely viewed piece on *Dateline NBC* took small beer to task, placing Boston Beer directly in the crosshairs. It questioned the beer's quality, highlighted the fact that Boston Beer did not have a brewery of its own (the beer was being made on a contract basis at larger breweries, including a Miller facility), and generally doubted whether small-batch beer could compare to that produced by the existing behemoths. Rewatching the segment today is painful not only due to some of the legitimate questions it raised, but also because of the tremendous bias showed by an "impartial" news program. (Beer still, to this day, gets treated with less respect in larger media outlets than other beverages, especially in cooking segments on some of the lighter news programs.)

Following the broadcast, Anheuser-Busch purchased radio ads that specifically questioned Sam Adams's credentials and the quality of their microbrews. Many consumers who might have considered switching from traditional beers to the upstarts stuck instead with their old predictables. If you were a microbrew drinker back then, society saw you as on the fringe—and not in a good way. Ordering a small-batch beer, even a Boston Lager, could cause snickers. The burgeoning renaissance took a big hit.

Forest fires can devastate a region, but they also often help regenerate growth in wooded or grassland areas. Soil is replenished, trees release seeds, and land that was in dire straits becomes lush again. What happened to the microbrew industry during that time similarly felt like a purge and renewal. Many ill-fated or poorly run breweries died by the wayside. The survivors left standing were a more vibrant, focused group of small-batch beer makers determined to stay committed to quality. They refocused on their messages, and revamped and dialed in high-quality recipes.

THIS MOMENT ALSO GAVE BIRTH TO THE WORD MOST COMMONLY used to describe small-batch brewing today: *craft*. While the term was born out of necessity—and a need to rebrand after the disaster of "microbrews"—it's my belief that these five letters have created nothing short of chaos for beer.

Although he was probably not the first to put the two words together, author Vince Cottone is credited, in the United States at least, with christening the industry "craft beer." In his 1986 book, *Good Beer Guide: Brewers and Pubs of the Pacific Northwest,* he used the phrase without firmly defining it, but the industry was so small that readers just understood what he meant.

These days it's a bit different. Defining craft beer is about as easy as nailing Jell-O to the wall. Still, many have sought to try. Is a craft beer one made with passion, out of pure and honest ingredients, and in the spirit of tradition? Is it made in small batches, and if so, "small" compared to what on a numbers level? Is craft beer just something you know when you see it? The answer to all three questions is yes, and maybe.

So what is a "craft beer" today, and what does the word really mean? This question, as often as it is posed, still lacks a clear answer that suits everyone. Ultimately *craft* is a word that attempts to highlight the difference between what some brewers and consumers see in the "us versus them" fight. The "us" is small, local, hands-on beer made in facilities where you can meet the brewers and see the brewing process up close. The "them" are large brewing companies like Anheuser-Busch InBev, Heineken, MillerCoors, and others, where consumer experiences are more theme park—manufactured events—than personal.

The Brewers Association, a trade group that represents small breweries in the United States, wants to recognize its members as "craft brewers" while being careful to never actually define the term *craft beer*. Today the Brewers Association classifies its members as "small," "traditional," or "independent." A "small" brewer makes less than six million barrels of beer per year (a quantity that was bumped up from two million barrels to accommodate growth by members like Boston Beer). "Traditional" means the majority of beer comes from "traditional or innovative" beer-making ingredients. "Independent" means the brewery is at least 75 percent controlled by a craft brewer.*

All three definitions have been modified over the years to help bring in new members, to keep existing ones in the fold, and to more sharply define the lines between members and what most small breweries see as their biggest threat: large brewing companies with deep pockets who use ruthless tactics to stomp out the little guys.

"Craft" is already a beer category that brings out strong passions and blind alliances. When you add other countries into the mix, and their different systems for defining the term, the definition gets even muddier. But over and over again it comes down to that "us versus

* Here's a closer look at those classifications. "Small": Beer production is attributed to the rules of alternating proprietorships. "Traditional": A brewer that has a majority of its total beverage alcohol volume in beers whose flavor derives from "traditional or innovative" brewing ingredients and their fermentation. Flavored malt beverages (FMBs) are not considered beers. "Independent": Less than 25 percent of the brewery is owned or controlled (or equivalent economic interest) by an alcohol-industry member that is not itself a craft brewer.

them" argument. In these long-running debates, a common phrase you'll encounter is "drink craft, not crap." This implies that simply a brewer's large size or robust corporate ownership somehow means that their liquid quality is less than exceptional.

Now, it's true that beers produced by large companies can fall on the generic side. Again, you wouldn't be wrong to think of beer-flavored beer. But even if that is the case, and drinking an American light lager brewed with rice isn't your first, or tenth, choice, it seems to me that mocking the big companies is actually unfair. We should never diminish the skill that goes into making tens of millions of barrels of the same beer each year, at multiple locations, and all without defect. These beers are pitch-perfect at what they aim to be: simple, clean, inoffensive lagers. This perfection is more than many smaller "craft" brewers are able to accomplish on a regular basis.

I was at a recent gathering of Florida small brewers where the speaker at the dais reminded the community that selling even one bad pint to a consumer would do more harm than all the marketing dollars the big breweries might throw around could ever do. I agreed wholeheartedly, but many brewers didn't, and it seemed to me that they failed to understand the severity of the situation, or at least weren't willing to acknowledge that they were part of the problem. Craft brewers and the fervent consumers who support them have been living within a pretty solid bubble for a long time now. They believe that what they are doing in brewing—regardless of their beer's quality—is the most important thing that has ever happened to beer.

I'll agree that it's remarkable that these days there are local breweries for the majority of Americans to call their own. This is astounding considering where we were just forty years ago. But if the beer certain companies are serving is full of defects, or simply striving for meh, does that do anyone any good? Consumers deserve better for their money than mediocre beer. And if brewers of lesser-quality beers don't step up their game, they'll eventually be forced to close from lack of business.

But "it's okay," I hear again and again. They're "craft." We need to "support craft brewing."

I disagree. We need to support good brewing and good brewing only.

Furthermore, the word *craft* itself has become useless. "Craft" has never actually been as black and white as "us versus them." And in recent years the situation has gotten even more complicated. Companies like Anheuser-Busch InBev started buying up smaller breweries and not always making corporate ownership clear on labels. This came to light during a Brewers Association public relations campaign dubbed "craft vs. crafty," which attacked brands they believed were hijacking words such as *craft* and *artisan* from their association members. The condemnation was aimed at beers like Blue Moon and Shock Top, whose corporate owners are MillerCoors and Anheuser-Busch InBev, respectively. It also lumped in breweries like Widmer Brothers in Oregon, part of the Craft Brew Alliance, which includes the brands Kona and Redhook, along with a handful of other, smaller breweries. A decade ago Craft Brew Alliance signed a distribution deal with "big beer" makers so that it could break into new markets. Because the smaller companies had given up partial ownership, the Brewers Association wanted to officially kick them out of the "craft" club.

But did the beer change, or the company's ethos? Not as far as I—or many others—can see. In fact, Portland is home to nearly one hundred breweries, most of them small, and nearly everyone in town will talk about how Widmer, the city's largest brewery, is a fine corporate neighbor, still producing high-quality beers, only now on a larger scale, and regularly helping smaller local breweries by providing lab access and other services.

But they are "crafty," according to the campaign. Beware.

THE THOUGHT BEHIND THE BREWERS ASSOCIATION "CRAFT VS. crafty" campaign was a good one, in that it encouraged consumers to be knowledgeable about brewery ownership, but the execution was wrong, and ultimately revealed some dated thinking. Some breweries—long-standing American companies like D. G. Yuengling and Son of Pottsville, Pennsylvania, and August Schell Brewing of New Ulm, Minnesota—were rather absurdly lumped into the "crafty" category based solely on ingredients.

In 2012, when the BA launched "craft vs. crafty," they made their distinctions based on the "three pillars of craft" mentioned above: small, traditional, and independent. It was the "traditional" element that landed Yuengling and Schell in the craft-drinkers' crosshairs. Both of those breweries had been around since before Prohibition. They survived the dark times and emerged stronger than ever, still making large volumes of beer. However, they used rice or corn in their recipes—and had always done so—which purists deemed nontraditional. This disdain for those ingredients didn't take into account that when the breweries originally opened (Pabst and Anheuser-Busch hail from a similar era), these adjuncts were the fermentable grains available in abundance. It's not that the brewers didn't want to use barley; it's that there wasn't enough of it to go around. So they needed to supplement. Yes, rice and corn are cost-effective ingredients, but they are also part of these companies' original recipes.

Jace Marti, the sixth-generation brewer at Schell's, wrote an impassioned open letter to the brewing community and its customers challenging the BA's definition of "craft brewery." He wrote, in part:

> For you to say that the three oldest, family-owned breweries in America are "not traditional" is downright disrespectful, rude and quite frankly, embarrassing. If you want to keep us on your list of shame, then so be it. That is your decision. We will continue to pour our heart and soul into every drop of beer that we make in this small, independent, and traditional brewery. Just like every other craft brewery out there does, and just like we have done for over a century and a half. Shame on you.

He was right. It was the awakening that many needed in order to realize that the voice from the mountaintop isn't always right or absolute. It was a turning point for many fellow brewers and drinkers. The BA backed off the campaign and changed its definitions to include breweries like Schell and Yuengling. This was not only a smart public relations move; it also allowed the association to add the output of those large heritage breweries to its annual volume as "craft" beer, thus increasing overall market share.

This moment in the BA's history was the beginning of the end for the word *craft,* because defining the term is impossible. It was also around this time that larger breweries began to purchase existing craft brands. In 2011 Anheuser-Busch InBev acquired Chicago's Goose Island Beer Company, a well-respected regional brewery that produced a respectable lineup of beers. Fans decried the loss of a beloved craft brewery to a conglomerate, and some shop owners spilled their existing inventory down the drain in protest. Yet despite the change in ownership, it would be hard to argue that their quality has gone down. (A notable exception was the 2015 release of the Goose Island Bourbon County Brand Stout. Some bottles needed to be recalled due to an unintentional *Lactobacillus* contamination.) True, some long-time fans say it just doesn't taste the same, and the reality is that the recipes for some of the beers have been modified. That's a natural evolution that occurs for most beers at many breweries, but among the fervent drinkers with an axe to grind it's just another reason to now hate Goose. However, for the countless drinkers who were introduced to the brand and its beers after the sale, it's unlikely they have nearly as much passion when the topic is brought up.

Since the purchase of Goose Island, Anheuser-Busch InBev has bought a number of other craft breweries to add to its high-end portfolio. These include Blue Point Brewing Company (New York), Breckenridge Brewery (Colorado), and Golden Road Brewing (California). MillerCoors has also gotten into the act with the purchase of Hop Valley (Oregon), Saint Archer (California), and others. Heineken has purchased Lagunitas Brewing Company (California). Japanese brewer Kirin owns 24.5 percent of the Brooklyn Brewery, which in turn has a stake in 21st Amendment Brewery (California) and Funkwerks Brewery (Colorado).

What about Brewery Ommegang (New York), Firestone Walker Brewing Company (California), and Boulevard Brewing Company (Missouri), which are now owned by the Belgian Duvel-Moortgat? Are they making inferior beer because they are no longer "American" owned? No.

Don't forget the private equity and venture capital firms that have been purchasing breweries over the last few years. Boston's Fireman Capital mostly owns Oskar Blues Brewery (Colorado), Cigar City Brewing (Florida), Perrin Brewing Company (Michigan),

Wasatch Brewery and Squatters Craft Beers (both in Utah), and Deep Ellum (Texas). Ulysses Capital purchased Victory Brewing (Pennsylvania) and Southern Tier Brewing (New York), built a new brewery in Charlotte, North Carolina, and is actively on the hunt for more breweries to add to the fold. Encore Consumer Capital of San Francisco bought out the employee-owners of Full Sail Brewing Company (Oregon) a few years ago. And the list goes on.

This blurring of the lines is happening because both sides, small-batch and mass-produced alike, are trying to steal consumers from the other. In this vein, becoming employee-owned is another tactic that some brewers are embracing. New Belgium Brewing Company, the fourth-largest craft brewery in the country, is employee-owned. The same is true for Harpoon Brewery (Boston), Deschutes Brewery (Oregon), New Glarus Brewing Company (Wisconsin), Modern Times Beer (California), and more. Many breweries that have opened within the last five years have offered employees equity in the companies. Not only does this practice help retain good talent and give employees a real incentive to work hard and grow the business, but for consumers it's the larger-scale brewery version of a mom-and-pop operation. Knowing that your money goes toward what's essentially a family-owned business, even if it has multistate distribution and produces hundreds of thousands of gallons of beer a year, feels good.

The word *craft* has been helpful as an organizational rallying cry for the Brewers Association, but in truth it is also about marketing. Now the so-called big guys are getting in on the game. They know there's a trend these days toward small and regionally specific products, and they want to capitalize on consumers' desire to buy local. That's why we've seen brands like Blue Moon (owned, as mentioned, by MillerCoors) using the phrase "Artfully Crafted" in their advertisements. And it's worth noting that Boston Beer, makers of Samuel Adams, used the line "For the Love of Beer" in its ads around the same time, with no mention of the word *craft.* It's part of a calculated move to pull in more customers from among the ranks of all beer drinkers, not just the ones who prefer "craft" brew.

One word shouldn't be a dividing point. Ultimately, what's important is the beer in the glass, and whether it tastes good to the individual drinker. In the same way that the term *microbrew* is still batted around, I don't honestly believe that the word *craft* will disappear anytime soon, but I do believe it's time to have a conversation about what it really means. Is it a helpful word that makes beer better? At first blush, it's business as usual for the breweries that have sold to other entities. They are making good beer, using fine ingredients, and innovating. Whether we call the result "craft" or not, should ownership matter?

THE SHORT ANSWER IS YES.

This book was written on a MacBook Pro on a program owned by Google and edited on a program owned by Microsoft. I drive a Ford, make calls on a Samsung phone, and listen to music through Bose headphones. My dress shirts are made by Tommy Hilfiger (because I like the fit). Dozens of items in my everyday life come from large, soulless corporations. That's not to say there aren't talented or passionate people among the company ranks, and that the companies weren't, once upon a time, start-ups themselves, but they are now profit-driven machines that have cornered a section of a market.

However, in other areas there remain other choices. We get our morning bagels from the baker down the street, not the supermarket. We buy books from the local independent bookshop, trying to avoid the Amazon juggernaut, and my bow ties come from a small company in Vermont, where each is hand sewn.

Beer is personal. It elicits thoughts and emotions in each drinker. We're at a remarkable point in the beverage's history where we can, once again, drink local. Where we can go and meet the brewer, have beers made with local ingredients, and share a pint with like-minded consumers. I'm still friends with Jeff, that South Orange bartender who poured me my first pint.

Ownership matters, because we still have that choice. If you're the kind of person who enjoys—or would like to enjoy—locally made beer, you'd also probably entertain the notion of buying a locally made computer, or cell phone, or car. But those options don't exist on a small scale, so we buy what's available.

With beer, there are still options, and knowing where your dollars go is important. Small doesn't necessarily mean good, but freedom of choice remains important.

From my viewpoint covering the industry, I don't need tea leaves to see the changes rapidly approaching. A slew of new breweries is opening. Existing brewers are growing. More mergers and sales and business closings will occur. There are going to be so many shades of gray in the conversation about "craft" versus "not-craft" that it will be difficult to differentiate one from the other. And this isn't even accounting for the international beer scene, which blows my whole argument to pieces.

In the past I regularly used the word *craft* without a strict definition. I even used it in the title of my 2013 *American Craft Beer Cookbook.* Since then I've done some soul searching about the word and what it truly means, outside of the trade association's definition. And I think the term is not long for this world. I believe that soon enough *craft* is going to join its brother *microbrew* as a word that once meant something specific but no longer does, a symbol of a period of time in the beer industry when folks were still figuring it all out.

While that happens, is it really necessary to further an "us versus them" mindset? There are, honestly, more important things to worry about.

During my time as the editor of *All About Beer Magazine,* we phased out the word *craft* in our coverage, and I will do the same in these pages. I'm not going to differentiate between "craft" beers and not, unless I'm quoting others or discussing a previously defined membership group or segment of the market.

ALTHOUGH THE PASSION THAT HAS SURROUNDED THE WORD *CRAFT* won't truly ever go away, there's a new word replacing it. The trend started slowly and is quickly gaining steam. With mergers and consolidations continuing, and Brink's trucks of cash stacking up at brewery doorsteps from venture capital firms, private equity investors, and larger brewing companies, smaller brewers looking to regain street cred with a certain sect of drinkers have turned to a new descriptor: *independent.*

In beer today, it's no longer big versus little—it's everyone for themselves. The fight for stock-keeping units (i.e., product-identification codes; abbreviated SKUs and pronounced "skews") and tap handles has led to skullduggery among smaller brewers. The growing millennial segment of the customer base, unlike generations before, is not loyal to just one type of liquid. A typical night out for them can mean starting with a beer, moving on to wine with dinner, and enjoying a cocktail afterward. A single consumer bringing other beverages into the mix means a tough road for the beer makers.

In this competitive market, after years of taking peashooter shots from David, the Goliaths are swinging back. They've already undermined the exclusivity of the word *craft*—and there's no denying that beers coming from corporate-owned brands like Goose Island (where people still line up in droves for that Bourbon County Brand Stout), or Elysian, or 10 Barrel, or Camden Town (in the UK) are quality products, packed with the flavors that most beer drinkers enjoy. So now, for the smaller breweries, it's not about the liquid; it's about the independence.

Claiming "independence" is tougher for the larger brewing companies to do, so it's the road the brewers formerly known as "craft" are walking. But keep in mind that "independence" means less than 25 percent outside ownership. That leaves a lot of room for outside money. I already mentioned that Brooklyn Brewery is *juuuust* under 25 percent owned by Kirin, and Dogfish Head sold 15 percent to LNK Partners for an undisclosed sum. These brewers all say that this kind of funding won't have an impact on the spirit or vision of the brewery, and they still tout their independence.

Op-eds have appeared in major newspapers about the importance of independent breweries. Even Tröegs, of Hershey, Pennsylvania, rebranded its label, packaging the company (at least the forward-facing part) as Tröegs Independent Brewing. In 2017 the Brewers Association released an official seal for their members (oddly it depicted an overturned bottle) to claim independence and help steer drinkers toward specific beers.

Independent. It's a solid-sounding word. It's what this country was founded on and as such resonates with every American who hears it, regardless of whether the topic is beer or something else. The ability to stand on your own two feet and follow your own path. It seems promising, but the jury is still out over whether using that word will help brewers sell more beer or get people to attend their events.

I believe that the description *independent* should reflect more than corporate structure. It has to be a guiding light. Brewers who use the seal and the word should really stop and think about what it means to them. Does it make your beer any better? Does it make us, the consumers, want to give up our money?

It's too early to say how the word will impact the industry, but any of us who use it, consumer or brewer, shouldn't get behind it just because we're trying to establish the terms of the next "us versus them" battle. We need to ask how being independent benefits the employees, the beer, the community, and all those supporting this movement.

It's easy to criticize the "big guys" and the "traitors" that are no longer "craft" or "independent." With each new sale there are gut-wrenching blog posts about the evils of Anheuser-Busch InBev and the companies they have made significant deals with (through an investment arm known as ZX Ventures), like Northern Brewer Homebrew Supply, RateBeer.com, PicoBrew homebrew kits, or hop farms across the world. Oddly, because they aren't the traditional enemy, there's less criticism when it's a venture capital firm cutting down a fellow brewer in order to make money and get a competitive edge.

While being part of a group and displaying an independent seal is important, so is standing on your own two feet. If you feel strongly about independence—if you're a consumer or a brewer or both—and want to air your criticisms, especially if they are founded, you need to be vocal, even if it means calling out other folks on the "us" side. The industry is still mostly asshole-free, but as with all industries, problems remain.

Increasingly, frustrations are leaking out online among brewers. This especially happens on social media, where the looks of a beer, an off flavor, underhanded practices, and other issues are often anonymously brought to light. It's easy to throw rocks at the big brewers, yes, but pointed yet polite criticisms should be levied under the commenter's real name to smaller entities as well. If you're a brewer who is going to complain about one brewery, then complain about *anyone* producing subpar product. People shouldn't get a pass just because they're in your group.

When we hold a glass of beer in our hand, it's the promise of good things to come, or an escape. It's the delight in tasting new flavors or teasing out nuance in familiar favorites. But beer is also a business, and a big one. It accounts for billions of dollars in innovation, marketing, and sales each year. It's the livelihood of many, the passion of even more, and the scourge of some.

Knowing where this industry has come from, especially here in the United States, should help us appreciate where we are today and the variety of possible futures. When we get bogged down in definitions and in-fighting, we're ultimately distracted from the lovely end result, and some of the joy is lost.

Personally, I'm a fan of local, of small, of the underdog. I like knowing that my money supports local businesses, family-owned companies, and ultimately fuels innovation and good flavor. Professionally, in this war of beer I'm a conscientious objector. I'll drink any beer, talk with any brewer, and cover the news of all breweries regardless of size or ownership.

At the bar, over pints, and certainly online I've seen people get worked up over just one word. Worrying about a definition that is written in sand takes away from the overall drinking experience. I don't worry about classifications. I just call it beer.

TWO

THE FOUR MAIN INGREDIENTS

For something so ubiquitous, and considered one of adulthood's great pleasures, it can be easy to forget all that goes into making a beer. It's made not only of the four main ingredients—water, malt, hops, and yeast—but also of a history that stretches back to the beginning of humanity. Throughout most of the world, beer is a part of the fabric that ties us all together.

There's a lot of fabric. The hundreds of millions of barrels of beer being made globally each year comprise more than 100 styles using all manner of ingredients, imparting new flavors and breaking new ground on a daily basis.

Because beer has been around for so long, it's easy to take it for granted. We know our ancestors drank it. We know there are different types, and if we've drunk a good variety, we know we prefer some styles over others. We know that by putting dollars down on the bar we're stimulating the economy. These are truths that even the most casual of beer drinkers accept. The beer in our glass is interesting in and of itself—and hopefully delicious—and when we're drinking, that's often all we think about: the sum of the parts. When we take a step back, however, and start considering the chemistry that led up to what we hold in our hands, beer becomes even more interesting. Knowledge is the great adventure, and beer is an excellent Sherpa. And it begins, of course, with the ingredients.

As mentioned, beer is composed of four primary ingredients: water, malt, hops, and yeast. It's been this way for centuries as brewers have strived to make pure, clear (usually) beer of uncompromising flavor that highlights the nuances and strengths of those humble components. What matters is how the four are mixed, tinkered with, and fermented. The end result can be the difference between a great brew and one not suitable to water plants.

Over time, and especially in the last two decades, brewers around the planet have begun to experiment with adding specialty ingredients to the main four. The results range from the sublime to the bizarre. Basically, if you can ingest it, you can brew with it. Included in these novel ingredients are coffee, tea, chocolate, fruit (of the traditional, tropical, and exotic varieties), garden herbs, spices, marijuana, wood, stones, mushrooms, meat, oysters and other shellfish, vegetables (e.g., peppers and cucumbers), salt, milk, and sugar. I've even come across animal-poop-smoked beers. As well as one made with Rocky Mountain oysters, another brewed with actual money, still others fermented on the bottom of the ocean, and one served in a bottle stuffed inside a taxidermied rodent.

Those last few are attention-getting for sure. Some unusual beers can be quite good, but usually they serve to shock, not satisfy. A beer that immediately comes to mind was made with grilled beef hearts and rosemary; Samuel Adams brewed it with Chef David Burke several years ago. Opening a 12-ounce bottle (which had a very short shelf life) released the metallic scent of blood into the air. Most folks only had a sip and dumped the rest of the bottle. Some ingredients are added to raise eyebrows while offering no discernable flavor, like when Dogfish Head Craft-Brewed Ales added moon dust to a pale ale. (Yes. Moon dust. More on that later in the book.)

Such beers are made with varying results, but what they highlight is that beer is a beverage that can be constantly tinkered with, improved upon. Or simply something someone did one time, never to repeat again. But I'm getting ahead of myself. Before we can truly appreciate where beer is now, with its unshackled acceptance of every ingredient possible, we should take some time to think about the four main ingredients, the foundation of every beer: their place in each recipe and indeed in the larger beer-drinking experience.

I've judged beer contests around the world, discussed pints with friends at the bar, and conducted tastings for both print and pleasure at my home. The term that drives me bonkers is *balanced.* It shows up more often than it should in blog posts, podcasts, and conversations. It's a lazy way to describe a beer (the same goes with *backbone* when it comes to describing malt, and *solid* when folks don't want to talk nuance), and even I am guilty of having used it in the past. What people usually mean when they say "balanced" is that the sweetness of the malt is not overpowered by the bitterness of the hops, so those two ingredients work well together, imparting desired flavors. I'm all for that in a beer. I like tasting individual ingredients, and that's a necessary part of the experience. But the descriptor *balanced* foolishly discounts the other two ingredients: water and yeast.

For a long time, one of the most celebrated and sought-after beers in the country has been Heady Topper, an IPA from the Alchemist Brewery in Stowe, Vermont. A beer ahead of its time, it was one of the original hazy New England IPAs, a cloudy, hop-aroma-forward ale. I believe one of the reasons the beer is so revered is because it deftly uses all four ingredients in complete cooperation with each other. I once wrote in a review for *All About Beer* that the ingredients fit together on the "head of a pin."

The more we focus on malt, and especially on hops, while ignoring the other two, the more we risk losing what makes beer special in the first place. In my opinion, this tendency to discount the importance of water and yeast comes down to the fact that for us casual drinkers who lack degrees in microbiology or agriculture, hops and malt are easier to grasp in our minds. They are not only tangible; we can taste and smell them individually.

I believe that looking at all the ingredients in a more practical than analytical way helps us understand each on its own, and then how they interact with each other. When that is achieved, we can truly appreciate beer for all it is, forging a deeper relationship with it.

WATER

WATER, OF COURSE, IS THE MAIN INGREDIENT IN BEER. ITS IMPORtance cannot be understated. Without clean, fresh water, there is no beer. Depending on where it is drawn from, it can be hard or soft. It can have a mineral-like flavor, even a tinge of salt. Varied water quality is important to different beer styles, and can be manipulated through technology to help a brewer reach a desired pH level. (Chemistry refresher: pH is the measure of how acidic or alkaline a substance is, with 7.0 indicating neutral. Pure water has a pH of 7.0, but most water isn't completely "pure," because it contains various minerals and dissolved gases. The lower the pH, the more acidic the substance, and the higher the pH, the more alkaline it is. Most of the water we drink has a pH between 6 and 8.5.)

Water is also, without a doubt, the ingredient most ignored by us the drinkers. Why should we care about something that's always readily available? This substance that, yes, we use to hydrate our bodies, but that we also use to wash our clothes and hair, flush our toilets, and is just there whenever we turn our faucet?

First of all, clean water isn't easy to come by. As we've seen in recent years, governmental entities whose job it is to keep water supplies safe sometimes fail. The water crisis in Flint, Michigan, brought to light the city's aging infrastructure and the general inability of governments to get in front of a public health problem. Environmental regulations are being rolled back. Lakes and rivers are routinely polluted, and the violators are rarely charged or, if they are, even more rarely punished, say environmental advocacy groups.

We all expect clean water—we all know that we *need* clean water. Yet the collective outrage we should all share when it gets mucked up never really materializes. This has especially been my observation when it comes to beer. Beer fans love to talk about hops and where they are grown, specific kinds of malt and their provenance, and even where yeast has been harvested (or which company provided the strain). It's rare to hear a drinker talk about water, and even more rare for people visiting breweries to ask about the local water source.

Before any new brewery is constructed, the first thing the entrepreneurs do is make sure there's access to a water source. Before electricity is hooked up, before the concrete slab is poured, before warehouse walls are built or stainless steel tanks moved in, there needs to be water. Without water, we'd just be chewing on the other ingredients.

Depending on where the brewery is located, brewers might be able to simply open the faucet and get down to brewing, so long as the water is complementary to the beer they want to make. Brewers appreciate more than anyone that water doesn't simply taste like water. The taste depends on the location. We can taste the difference between city water and well water from a country pump that has a light cool taste. Water in shore regions can have a salinity content. Depending on a brewery's site, these flavors can come through in the overall taste of a beer.

As mentioned, different styles of beer benefit from different kinds of water. Stouts can benefit from water with high alkalinity (a higher pH); hard water benefits darker lagers; low-mineral water is favored by pilsners. Perhaps the most famous brewery town celebrated for its water is Burton upon Trent, in the United Kingdom. It has been home to a multitude of breweries over the centuries, with Bass perhaps being the best known today. The water is high in calcium and magnesium, low in bicarbonate and sodium with a distinct sulphur aroma. It proved lovely for English pale ales, giving them a distinct taste sought after all over the world.

Today, if a brewery in, say, Phoenix wanted to make a traditional English pale, with the right brewhouse technology they would be able to bring their local water to a neutral pH, then add various mineral compounds to re-create the flavor and aroma of water found in that central English town. It's the same general method used by large brewers—with plants all over the country, if not all over the world—to make tens of millions of barrels of beer each year that taste the same. Budweiser, the iconic American lager, is produced in a dozen breweries scattered across the United States. The brewery's reputation depends on each batch tasting exactly like the last one, which means customers should be unable to differentiate one beer from another based on location. In the same way that all the other raw ingredients in a bottle of Bud come from the same growing locations and are of the same quality, by controlling their water and manipulating it to their needs, the Bud made in Newark, New Jersey, tastes just like the one coming out of Fairfield, California. Whether you like Bud or not, this is a technical feat of mastery worth appreciating.

Brewers know that without clean water they are out of a job. It's one of the reasons why brewers like John Kimmich of the Alchemist in Vermont, and Jaime Jurado, a master brewer who was worked for breweries like Abita and Gambrinus, are two of the hundreds of brewers I've visited over the years who were eager to show me their water-filtration and wastewater systems before we even examined the brewhouse, which is usually considered the shining jewel of any tour.

Water use goes both ways. The brew day begins and ends with water, and there's a lot of it used in between, but there's also a fair amount left over. Brewers obviously want the water that comes in to be in good condition, but many also treat their water before it leaves. Sanitation is key at a brewery. It keeps infections out of the beer, and, because beer is a food product, brewers are subject to health inspections and other sanitary standards. A lot of cleaning chemicals and caustic substances are used to keep their equipment clean. Brewers are, by and large, environmentally conscious and will either use substances that are effective but less harmful, or have reservoirs or basins in place to treat the cleaning water before it is released back to either the municipal sewer system or the ground.

Many brewers are also activists. A coalition in the Great Lakes supports keeping those majestic bodies of water as pristine as possible. The lakes are not only a huge economic driver for those regions, but yup, you guessed it: some are the source of the brewery water. Brewers in Alaska have been mindful of oil drilling, and they worry for parts of the plains where hydraulic fracking is becoming more common.

Vaguely reminiscent of the days of the American Revolution, when taverns became the places where plans were formed and action fomented, breweries sometimes use their taprooms to talk with customers about the importance of clean water. Peaceful activism there can have not only a local impact, but a national one too. And it's not just talk. Some breweries allow science and technology companies to test new ideas and equipment with the goal of cleaner water.

Take, for example, the half dozen Boston-area breweries that participated in a competition to make a beer brewed with water from the Charles River, the landmark waterway that was once so polluted that swimming was restricted. A Massachusetts company pulled four thousand gallons from the river, and through reverse osmosis purified it and sent it to local brewers. The results varied from pale ales to porters, and the ultimate reward was the ability to prove to consumers that some water can be reclaimed, and that we all need to think both about preventing pollution and about finding ways to fix existing problems. You couldn't tell that river water was used to brew the beers, and most gave the taste a thumbs up.

We each have a responsibility to treat our water sources with respect, to demand access to unadulterated drinking water for every citizen of the planet, and to make sure companies who try to skirt regulations are held accountable.

MALT

Technically this section should be called "Grains." Although malted barley, specifically the two-row variety, is the most common type of brewing grain used, it's not the only kind. Rice, corn, rye, wheat, and certain other grains may also be used in the brewing process (although in some circles *rice* is a dirty word). Some of them are suitable for use in gluten-free beers.

For the most part, the grain used in beer comes from Canada and America's breadbasket: Idaho, Montana, and the Dakotas, where farms stretch for miles and grains grow high and sway in the winds. It's a captivating and awe-inspiring sight, and a source of national pride. As schoolchildren we sang about the "amber waves of grain" right alongside our nation's "purple-mountain majesties" and "fruited plains." The United States grows tens of millions of bushels of grain per year, according to the Department of Agriculture, the majority of which goes to food production, livestock feed, and other purposes. But some (a smaller percentage than you might think) gets turned into beer. That pint of pale ale is downright patriotic.

Malt's main purpose is to provide the sugars necessary to make beer. Malting is the hydration, germination, and drying of the grain to prepare it for the brewing process. After it is harvested and brought to a malting facility, the grain is soaked and each kernel begins the germinating process where the breakdown of proteins and starches begins. Next, the sprouted grain, known as chit, is spread out onto large sheets where it is dosed with humidified air and rotated, allowing the starch reserves of grain to open up as the germination process continues. Then the grain is transferred to the drying phase where it is kilned, or roasted, to bring out additional flavors. For traditional brewers, like Coors, malting is done on an almost mind-bogglingly large scale at hulking plants that were built to suit the needs of companies that use tons and tons of raw ingredients to make their beer.

A simple pale malt can go through extraordinary transformations during the kilning process. Depending on how it is kilned, malt starts off tasting a lot like plain cereal. The more it's roasted the more complex it becomes, taking on the flavors of Grape-Nuts, caramel, chocolate, coffee, or toffee. If malt burns it can taste like acrid black carbon, and although that might sound less than ideal, think of a wood-fired pizza fresh from the oven. It's those black, burned bits on the crust that many of us love to nibble first. All these flavors are familiar; we encounter them in everyday life, from breakfast to dessert. We are so accustomed to them that they are easy to ignore, but each one should be noticed and celebrated within each pint.

Malt "is the fair-headed stepchild" of brewing, the brewer Jaime Jurado once told me. Despite all it does for the appearance, aroma, and flavor of a finished beer, and how vital it is to the actual creation of beer, the ingredient itself is often overlooked in favor of hops. Hops are sexy. Hops are bursting with aroma. Fresh hops are photogenic, and they have such a fanatical following that you can quickly lose count if you try to tally the hop tattoos at a beer festival. On chalkboards at breweries across the country, brewers proudly announce the hop varieties used in various recipes. Chinook! Mosaic! Idaho 7! An experimental variety fresh from the farm! Malt rarely, if ever, gets a public mention. But hops can only remain the main focus for so long, and thankfully malt is starting to get its due.

Because of the aforementioned familiarity of the tastes and fragrances imparted by malt to the brew, it has a more subtle impact on a beer's flavor than hops do. But malt is the soul of beer. The majority of the beer's components come from malt, such as the alcohol and mouthfeel contributors. Some brewers say that thanks to the broad palette of colors, flavors, and styles associated with malt, there is still a lot of white space left to explore when it comes to creating new recipes and styles. Whereas hops inspire specific descriptors for tastes and aromas, the grain is often given only a passing notice. When I first started drinking beer, I'd use bland descriptors to portray malt. In more recent years I've tried to be as specific as possible about malt when tasting or judging a beer, because teasing out malt flavors is a difficult but rewarding experience.

A good number of brewers, without giving the grain too much thought, select their malt from a catalogue based on the recipe(s) they want to make. Others have dived deep into malting. They dedicate resources to finding links between the traits in different barley varieties, the malting processes, and the resulting malty flavor and color in the beer that will make their recipes pop. Some breweries, like Black Shirt in Denver, Colorado, make only red ales (using different ingredients in various recipes) to showcase the diversity and endless interest afforded by a malt-forward style. Such practices mean that the malt conversation is shifting. New suppliers of malt are opening, or breweries are installing their own malting facilities, all of which leads to increased consumer awareness and a more prominent role for malt in barroom conversations.

In the United States, thanks to the modern beer movement, the renewed interest in malt is occurring in no small part because many brewers are branching out from the traditional two-row barley that dominated recipes for decades. Six-row barley had been the dominant brewers crop until about the 1950s when American brewers started breeding two-row to function like six-row. The main difference between them is two-row has two kernels along the barley spike where six-row, as you'd imagine has six. Still, everything old is new again and it's not uncommon to find beers today brewed with six-row. These days, anything that can contribute extra depth or flavor to a beer is considered a suitable additive during the kilning process. One trend is for brewers to smoke their malts with other ingredients, like pork (imparting a bacon note to the finished beer), or using specialty woods such as apple or maple. Alaskan Brewing Company uses a salmon-smoking facility for their famed smoked porter, which adds an almost oily character to the finished beer.

Brewers are reaching out of their comfort zones in their efforts to produce beers with creative ingredients that will stand out in a crowded market. To source malt, they are looking beyond the traditional sources like Germany, the United Kingdom, Canada, and the United States to Scandinavia, Chile, Japan, and other regions. Because malting companies provide a crucial ingredient from a potentially far-flung locale, they are working hard to make sure their brewing customers know as much about malt products as possible. For instance, Japanese brewer Sapporo obtains its grain from the agriculture conglomerate Cargill. The two work together to track the practices of the farmers who supply barley for Sapporo's beers, logging detailed records regarding seeding, fertilization, farming practices, and harvest. Technology like this, as more brewers adopt it, will give brewers and beer lovers alike a better understanding of not only the ingredient itself, but also where it came from and the farmers who grow it. Brewers and growers are excited to share this information. Even for something as familiar as grains, the majority of us drinkers don't always appreciate the process—or people—who get it into our glass.

Of course, not every brewer chooses to operate at this industrial level. As the brewing industry has grown, niche businesses have popped up to support beer on a more local scale. Over the last decade we've seen the rise of the micro-maltsters. Micro-malting was developed to meet the needs of smaller neighborhood breweries that turn out only a few hundred or thousand barrels of beer per year. Micro-maltsters are small companies with relatively modest fields, like Valley Malt in Massachusetts, Admiral Maltings in California, Anson Mills in South Carolina, Rabbit Hill Malt in New Jersey, and the Colorado Malting Company. They supply specialty grain in small quantities to small, usually local breweries. For some breweries it's about using locally sourced ingredients—something they can tout to their consumers. For others it's about supporting a fellow small business.

Then there are the midsize breweries—smaller than the big players, bigger than the small ones—like Rogue Ales of Oregon and Bell's Brewery in Michigan. They farm their own barley fields and have built malt houses for germination and processing. Although not every brewery will construct a full malt house (a venture that requires a large staff commitment), an increasing number of breweries install their own malting machines. Typically, these are brewers that make more than fifty thousand barrels per year. Having malting equipment onsite allows a brewer to kiln grain to exact specifications and exercise complete control over their recipes. When a brewer makes that commitment, it's nearly impossible *not* to notice the results in the glass. A beer made with such care given to the malting process pops with flavor. It's like the difference between the taste of bread purchased at the grocery store and the loaf you make at home.

Once malt has served its purpose to provide sugar for the fermentation process intrinsic to beer-making, that doesn't mean it's no longer useful. Visit a brewery on a brew day and you'll see hundreds, if not thousands, of pounds of wet grain (termed "spent grain") being shoveled—likely still steaming—into containers. It can, and often does, go into landfills, but it is also regularly carted off to farms to be used as rather tasty animal feed. It has the consistency—although a little more fibrous and chewy—of oatmeal. It's perfectly fine to eat; standard two-row malt has a flavor close to that of cooked cereal. Pigs, cows, sheep, goats, and other barnyard animals love the stuff. Not only is reusing spent grain an economically wise decision—saving good foodstuffs from being wasted—but in many cases if an animal is being raised nearby for slaughter, there's a likelihood that the meat will wind up on the brewpub menu. Burger from a spent-grain-fed cow is a harmonious accompaniment to the pint next to it.

In recent years, the use of spent grain to feed animals has gone beyond the farm and into the pond. Aquaculture is quickly becoming an intriguing and important way of growing vegetables and raising fish for consumption. The general idea is that pools filled with fish (e.g., tilapia) are natural beds for all manner of plants, from lettuce to herbs like basil. The fish fertilize the plants through natural waste production, and the plants, in turn, clean the water for the fish. Spent grain is sometimes used to feed these farmed fish. One such operation is at West Sixth Brewing in Lexington, Kentucky, where an aquaculture farm is in full swing. Spent grain is transferred from the brewery to the fish tanks. The plants grow, and when the fish are big enough they wind up on the plate at the restaurant, along with the locally harvested salad. The business also supplies food to local charities and food banks.

A lot of breweries, already dog friendly, have taken to making dog treats out of spent grain mixed with a little peanut butter or another savory treat. We humans can benefit from spent grain too. Brewpubs routinely use it in pizza crusts, sandwich bread, and malted milkshakes. You can find easy recipes—including two in my *American Craft Beer Cookbook*—for making spent-grain bread at home. Some breweries keep spent grain stocked in jars for enterprising customers (or you can bring your own jar). And because you're doing the business a favor by taking it off their hands, they may not even charge you for it—although it's always polite to tip a few extra bucks for the privilege.

As more brewers turn their focus to showcasing this ingredient, we drinkers are going to have a lot more to talk about.

HOPS

ONCE YOU LEARN ABOUT HOPS, IT'S HARD NOT TO LOVE THEM. It's not because they are part of the cannabis family—they are, but you can't do with hops what you do with cannabis. Honest. Don't try. It won't end well. Hops are good for one thing only, and that's making beer. Some companies have put hop leaves into soap for exfoliating purposes, and others use the oil in hand creams, but frankly that's just a waste of good hops that could have otherwise wound up in beer.

You might think of hops as buds, but they are actually small strobiles that grow vertically on bines, or long stems. They come from a perennial plant that needs a specific environment to grow cones; it thrives best between the 50th and 40th parallels, but can grow as low as the 30th parallel (in both hemispheres). Hops add bitterness, aroma, and flavor to beer.

The best-known hop-growing countries are the United States and the United Kingdom, plus the Czech Republic and Germany, from which come the four traditional hop varieties, known as "noble" hops: Hallertau, Saaz, Spalt, and Tettnanger. The flavors and aromas of these hops are generally described as spicy, earthy, or floral. They have been cultivated and used for centuries in pilsners and other classic beer styles that are specific to certain regions. Saaz, for example, is the noble-hop variety you should expect to taste when drinking a traditional Czech pils, like Pilsner Urquell. Spalt is common in a helles lager. Hallertau, sometimes called Hallertauer Mittelfrüh, is a common lager hop noted for its spicy character. You've likely had it several times. It's the dominant hop in Samuel Adams Boston Lager.

You've probably also been exposed to lovely aromas and flavors from United Kingdom hops, like East Kent Goldings (floral, spicy, honey), Fuggle (minty, earthy), First Gold (marmalade), and more.

These hops, and dozens more grown in that country, continue to be used on both sides of the pond and well beyond.

Over the last few decades, hops from the southern hemisphere, specifically Australian and New Zealand varieties, have burst onto the global market. Varieties like Nelson Sauvin and Wakatu have captured the attention of brewers and hop-loving drinkers (often referred to lovingly as "hop heads"). Although some wonderful tropical fruit aromas come from southern-hemisphere hops, such as pineapple, guava, passion fruit, and papaya, other characteristics that the Kiwis love might seem odd to us, like the faint whiff of petrol, or wet laundry.

In the United States, the top hop-growing region is the Pacific Northwest, which is responsible for nearly 97 percent of all hops harvested annually in this country, according to the Hop Growers of America. The climate in Idaho, Oregon, and Washington is ideal for hops, thanks to the long, hot summer days and cold winters. Hops grow best when the roots are wet and the leaves dry (bines can climb up to a foot per day in peak season). When someone refers to "US hops," they likely mean hops grown in that part of the country. An entire beer-tourism industry has sprung up in the region, with tours, museums, and immersive experiences designed to help consumers better understand the most beloved ingredient in beer. Although hop production in these states has slowed recently, they spent several years ramping it up, with growers innovating through breeding programs that yielded new flavors and more robust crops, and other states are following suit.

A quick bit of history: Prohibition damaged the American hop crop. Around that time, the mid-Atlantic and New England regions, particularly New York state, were home to hop farms that provided the hops to local breweries. The farms mostly disappeared during

Prohibition and never came back, although in some areas along the Saint Lawrence River you can still see tall stone towers that were once used to dry hop bines. During Prohibition hop growers in the Pacific Northwest exported their crop and then thrived following the repeal of the Eighteenth Amendment. These days, hop farms seem to be popping up in nearly every continental state, including Michigan and Florida. It seems odd to think of Florida as home to hop farms, because it lacks the right climate for traditional hop production, but that's what people said about blueberries. In the 1950s the University of Florida Institute of Food and Agricultural Sciences was tasked with finding a way to bring the berry to the Sunshine State; today it consistently ranks in the top five of blueberry-producing states, according to the Department of Agriculture.

One additional aside: another crop that was hit hard by Prohibition was apples, which are also making a comeback. People assumed that Prohibition would be the law of the land forever, so varieties of apple that were good for only one thing—being pressed for hard cider—were destroyed and replaced with culinary varieties that still thrive today. Over the last decade we've seen a resurgence of hard cider, which often appears on tap alongside beer. Several beer makers have gotten into the cider game, but when cider is produced on a large scale, as it too often is, culinary apples and concentrate—think apple juice—are often used instead of cider-specific apples. Some of the artisanal cider makers are working with institutions like the agricultural station at Cornell University, which cataloged apple cuttings before they were destroyed, to bring traditional apple varieties back. However, such strains typically have a low yield, meaning that any cider produced from them will be limited in quantity and hard to find. A prevailing hypothesis a few years back held that cider would become the new beer. Some producers even began hopping their ciders to appeal to beer drinkers. But the big cider boom has yet to materialize, and cider will likely stay on the margins.

THERE'S LITTLE SLOWING DOWN HOP PRODUCTION, OR THE EMPHASIS on hop-forward beers. IPA is the best-selling beer category in the "craft" segment (as defined by the research firm IRI); you'd be hard pressed to find a brewery that doesn't make one. Growers are struggling to keep up with demand and to create new varieties. Each year at hop harvest—typically mid-August through mid-September—brewers from around the country travel to the Pacific Northwest, notably Washington's Yakima Valley, to sample the year's harvest, place orders for the coming year, perform rubbings (opening the hop cones in your hands and grinding them together to release oils and aromas), and select lots they have contracted for.

The majority of the hops are dried, vacuum-packed (after pelletizing), and sent off to cold storage to await use in beer. A small amount of the crop is packaged fresh, or "wet," and sent to breweries within twenty-four hours of picking to be immediately added to a beer. These are called "fresh-hop" or "wet-hop" beers. They are usually IPAs, and they are bursting with fresh, bright, vibrant aromas.

If you pull a hop cone off the bine and split it open, you'll find a sticky, oily, yet powdery substance called lupulin—the source of most of hops' aromas and flavors. Lupulin powder is also used separately as an ingredient in beer, blended into the batch much like you would incorporate flour into a roux when cooking. Producers are also creating hop-extract oils, potent substances that only require a little to generate robust effects, and so flavorful that adding them to beer can mimic the taste of fresh-hopped beers any time of year.

Many scholarly books have been written on hops: on the history, the science, the breeding, the growing. It's impossible to get deeply into such topics here, but when you sit at a bar or in a brewery you're likely to encounter hop heads who go on and on about their love of hops, arguing over which varieties are superior to others and gossiping about the next "it" variety. Getting caught up in those conversations can be intimidating and exhausting. But they're also exciting because they perfectly highlight how far we've come as beer drinkers in America.

The descriptor that has long been used to describe the flavor imparted by hops to beer is *bitter*. It conjures in our lizard brains something unpleasant, a flavor to be avoided and replaced with one that's more pleasing, like sweetness. But think about it this way: you can't fully appreciate the sweet without the bitter, and although hops do have a hint of bitterness, it's unfair to paint them with such a broad and bold brush.

Hops are like the spice of beer and, depending on the variety, can impart fragrances and flavors with which we have positive associations. The most common flavor notes in hops are floral citrus: grapefruit, orange, lemon, and lime. Other varieties contribute pine or resin characteristics, more pleasantly earthy than the scents used in kitchen cleaning products. Some hop varieties offer hints of tropical fruits like mango and kiwi; others smell like honeydew melons, green onions, or even cheese. Newer strains impart aromas of strawberry, blueberry, or peach, or even subtypes of citrus, like mandarin orange and Meyer lemon. Sometimes hops can smell "catty"—an odor familiar to anyone who regularly changes a litter box. And yes, there are hop varieties that smell like marijuana, an aroma that brewers refer to as "dank."

Most of these flavors and fragrances are familiar to us. We enjoy grapefruit and oranges for breakfast, slice onions into our salads, and in certain states where it's legal we might partake in recreational cannabis use. When I encounter folks who swear they "don't like hops," or don't like beer because it's too bitter, I'll introduce them to an American pale ale, like the one made by Sierra Nevada. First I point out the familiar aromas and flavors, encouraging them to really smell the beer before sipping it—to tease out the pine, grapefruit, and floral notes imparted by the Cascade hops used in the beer. Once they've identified at least one recognizable fragrance, the hop experience becomes a little easier for them to embrace—and their enjoyment can blossom.

THE CURRENT AMERICAN BEER RENAISSANCE WAS BUILT ON HOPS. In the old days of beer-flavored beer, the large breweries used hops only in very small quantities. In many cases the hoppy notes were merely a whisper of a suggestion, even when Miller used marketing jargon like "triple-hop brewed" to sell its Lite beer. In response, brewers of the modern generation worked to boost the International Bittering Units (IBUs) in their beers, and for a while the thinking was "the more hops the better." Breweries ran in the opposite direction from the bland recipes that had ruled the land and instead threw full bines of hops at their customers until they submitted to the power of lupulin.

The strategy worked. Today, hops govern the conversation, a phenomenon that is both good and bad. Good because we get to drink flavorful beer, bad because so many hop heads are too quick to dismiss or argue with people who have yet to fully embrace the flavor.

I've witnessed these one-sided conversations more times than I can count. It usually starts with, "What do you mean you don't like hops?!" As though they are galled that someone has yet to learn a secret handshake. And rather than helpfully trying to ease them into the ingredient (in the way that I suspect they were once eased into it, and the way that I certainly was in my earliest days of drinking), they get exasperated and simply move on. This kind of hops-based interaction does no one any favors.

What's funny to me is that for all the hop love today, it wasn't always part of the conversation, or even part of the original beer recipe. There is chemical evidence for hopped beer in Europe dating back to around 500 BCE. But for centuries, all manner of herbs were used to flavor beer. The combination was known as gruit, and today that word is often used to describe a beer that does not contain hops. Some brewers still make the style, although less frequently. People want hops. Give them what they want.

Hops are added to beer at multiple points in the brewing process. In most recipes it happens during the boiling of the wort (the sugary liquid that results from boiling grain). There is also the practice of dry hopping, in which hops are added to the fermentation tanks, or even the serving kegs and casks, to impart a more vibrant hop experience. This practice, long in use by brewers, has increased in recent years with the advent of the substyle known as New England–style IPA, an intentionally hazy pale ale. Many brewers of the style add the letters "DDH" to their packaging and descriptions of the beer, to signify "double dry hopped." DDH is a topic that comes up regularly on my podcast *Steal This Beer* (mostly because it irks my cohost, Augie Carton). From his brewer's perspective: "Double dry hopped started being used to qualify an aromatic purpose of multiple dry hopping in the days of the bitter arms race. As the focus moved to aromatic potential the DDH short hand was co-opted. Whether the D for doubling meant number of additions or amount of addition stopped mattering as much as the message that there was an aromatic focus to the beer. Much like literally no longer means literally, DDH no longer means doubling or a second time but rather a process that moves a beer towards a fuller aroma experience and away from a bitter one." That's why beer labeled "DDH" is usually bursting with hop aroma and looks the part of a hazy, yellowish milkshake. And putting those letters on the packaging allows brewers to sell the beer for a few dollars more. Some folks even advocate that DDH should be its own separate sub-sub-style.

We've long since come to accept hops as an elemental part of beer, but it occurred to me several years ago that it was really a stroke of good fortune that they caught on as an ingredient. I remember having a conversation (over beers, of course) with Matt Kirkegaard, an Australian journalist who covers the beer industry down under. He asked me the following question, quite seriously: if beer were invented today for the first time, what would it look and taste like? After a little thought I came to the conclusion that it would probably not include hops. Since hops are used only in beer production and serve virtually no other purpose, I doubt that the plant would have survived over time, or been as widely cultivated.

To back me up, I later asked Stan Hieronymus, a journalist and author of *For the Love of Hops: The Practical Guide to Aroma, Bitterness, and the Culture of Hops*, for his thoughts on the matter. His reply:

> I agree that if beer were starting at year zero today, just like year zero the first time around, it would not contain hops. But hops still grow in the wild today—both in the US and in Europe—so the question would be how long would it take for somebody to put hops in beer, and then for somebody to discover the value of boiling hops. We don't know how the heck that happened first time around, so it would be a guess. But I suspect the understanding of isomerization through boiling would come quicker.

(Isomerization is when one type of molecule is transformed into another, with the same atoms but in a different arrangement.)

For his part, Kirkegaard believes that if something called "beer" were created today, it would be akin to a light lager, the kind that is typically pale yellow and served on beaches. A beverage designed not to offend, but rather to appeal to the largest number of people from all walks of life. We had such a lively conversation on these theoreticals that I opened up the discussion to other writers and thoughtful people in and around the brewing industry. They did not disappoint.

Carla Jean Lauter, a writer based in Portland, Maine, wrote, "I can see a few scenarios that would create a new genesis moment for beer. If we assume that everything else in our society remained the same and we are currently in the same situation—minus beer—I could see a few origin points." On the hops question, she said that brewers would likely experiment "with all kinds of bittering agents, until they reached hops. However, I think they'd be more likely to run through everything fermentable as an additive first—meaning that the first beers might be sours and fruit-laced."

Gerard Walen, author of the book *Florida Breweries,* agreed. "I wonder if the inventor would even see the need for a bittering agent. The collective palate of humanity has become accustomed to overwhelming sweetness as something pleasurable and desired."

The conversation on social media was a freewheeling affair, but ultimately there was general agreement that the trend would be toward sweet beers heavy on the fruit flavors. Some theorized that a stronger, malt-forward beer would be favored. Don Tse, a journalist from Canada, jumped into the fray and opined that the drink would be much more rooted in corn than barley.

> Corn is the dominant crop today and is used in everything (high-fructose corn syrup, for example, is in a lot of our food). Yes, I realize that corn is used in a lot of beer today, but I mean

> beer would be MUCH more corn-based. And if you look at [the] pre-hop days of beer, beer was spiced with all manner of herbs and spices, so I think that would be the case if beer were invented today. With corn being a lighter flavour, I think what would differentiate brands of beers would be the spices they use (not unlike how craft gin makers are now trying to differentiate themselves using different botanicals).

Solid logic even now, four years after that discussion took place.

For many others who weighed in, the concept of a recently invented beer was bleak. It would contain "caffeine, ginseng, taurine, guarana, and yellow dye number five," speculated Jonathan Moxey, a brewer at Rockwell Beer Co. in Missouri. Journalist Jeff Cioletti thinks it would be nothing more than fermented high-fructose corn syrup spiced with elderflower. That said, according to Kirkegaard, "As a new product it would need to be exciting enough to arouse interest."

Arouse interest, eh? With that criterion in mind, it's hard to imagine that brewers would never find their way back to hops, no matter when it was introduced.

YEAST

YEAST IS THE WONDERFUL MICROBE THAT MAKES BEER POSSIBLE. Different styles of beer require different yeast strains, which can deliver a multitude of flavors ranging from banana, bubblegum, and flowers, to pepper, clove, and funk.

A common saying is that brewers make wort, but only yeast can make beer. This is true: when yeast, a type of fungus, is added to the cooled wort (usually at the end of the brew day), the yeast essentially eats the wort, processing the blend into alcohol and leaving carbonation in its wake. Before science was able to explain this process, our long-ago ancestors believed that fermentation was a gift from the gods. What we know now is that yeast is indeed miraculous. Without it, we wouldn't have beer in the pint in front of us.

Yeast's role in beer—from sheer creation to the imparting of aromas and flavors—cannot be overstated. Some brewers highlight this influence by making a single style of beer but pitching different batches with different strains of yeast. When this is done well, you can still taste the base beer, but so much changes from glass to glass thanks to the yeast that the results are mind-blowing. Professional brewers will pull this stunt from time to time to shake up their tried and true recipes. Brewers can get into a groove when they use the same yeasts repeatedly; switching strains can result in pleasant surprises, allowing brewers and their customers to reengage in a new way with an old favorite beverage.

Beer falls into two overall styles: ales and lagers. Yeast is the key difference here. In the case of lagers, during fermentation the yeast gathers at the bottom of the tank. With ales, it gathers toward the top. (Other factors that differ between the two styles are timing and temperature. Ales can age for just a few weeks and prefer higher temperatures—in the 60s and 70s Fahrenheit—while lagers can age for weeks or months at temperatures between about freezing and 55 degrees. This is generally speaking of course; each recipe and strain of yeast is different.)

"Each type of yeast makes its best flavors at different temperatures," brewing instructor Steve Parkes told me. "Lager strains make good beer at colder temperatures, while ale yeasts make good beer at warmer temperatures." Ale yeast, *Saccharomyces cerevisiae,* commonly known as "brewer's yeast," usually produces fruity, spicy, or earthy flavors. Ales are the traditional beers of England and of Belgium and have fueled the modern beer renaissance here in America. This is because they offer familiar flavors and can be a little more palatable than other styles. It's also because ales are largely forgiving of flaws, like off flavors, which we'll discuss shortly.

When breweries and brewpubs began opening across the United States in the 1980s and 1990s (and even today), most stuck largely to ales. Ales mature in as little as three to ten days, meaning breweries could sell their finished beer much more quickly and turn around their tanks faster than if they were making lagers, which can take twenty-eight to forty days (as a general rule). Also, as the generation of new brewers forged a path, they needed recipes that could take a beating and possibly hide (or at least lessen) undesirable characteristics that might have come out of the brewing process. The brewers who were never able to overcome these problems were mostly left by the wayside. The ones who improved dialed in their ales to perfection, and many then began trying their hand at lagers, a much more exacting beer.

Beers in the lager family are fermented by yeast strains that, as mentioned, operate better at cooler temperatures (e.g., *Saccharomyces pastorianus,* also known as *Saccharomyces carlsbergensis*). These beers need to be conditioned or cellared (*lager* is German for "storage place") for an extended period of time to reach peak drinkability. The lagers are the traditional beers of Germany, the Czech Republic, and other countries of central Europe, and they are the dominant style here in the United States, anchored by Budweiser, Miller, and Coors. Incredibly fickle, lagers are difficult to make well. Regardless of a brewery's size, if they can make a technically perfect, defect-free lager or pilsner, they are worthy of applause.

Within these two main families, hundreds of yeast strains have been grabbed from the air, identified, catalogued, reproduced, and banked. Yeast banks are some of the coolest places you'll ever visit—think of the cryogenics lab in *Jurassic Park.* They range from nondescript buildings at Budweiser, in St. Louis, to the modern digs of White Labs, in San Diego and Asheville.

YEAST'S MODERN USE IN BEER IS A TESTAMENT TO HUMANS' DETERmination—or, really, our control over nature. Over the centuries, brewers have cultivated yeast strains to get them to a consistency that is most suitable for beer. Brewers essentially domesticated these yeasts, which is extremely unusual for a microorganism. Depending on the strain, brewer's yeast imparts some enjoyable flavors (like clove in a hefeweizen) that were selected over the centuries as brewers worked to perfect batches of beer. Brewer's yeast ferments only certain kinds of sugars—leaving some behind in the beer—whereas other strains consume all the sugars present in the wort. Brewer's yeast then settles in the fermenting tanks when it has finished its job, which is unique and not a naturally occurring trait. That's a bonus for bottle-conditioned beer, and for some homebrewers too. To save a buck, thrifty DIY-ers are getting creative, preserving and repropagating yeast found at the bottom of certain commercially available beers, and then using it in their own personal batches.

There are specific strains of yeast for every style of beer. Brewers only need flip through a catalogue and find the yeast that matches the recipe they want to make, from Vienna lager to hefeweizen to English porter. Certain breweries have worked to cultivate their own proprietary "house yeast," to give their beers unique flavors and characters that can't be found elsewhere.

The same is true for spontaneously fermented beers. Let's take the gueuze style as an example. After the brewers have made wort, it's pumped into a coolship (basically a large open pool) that is exposed to the elements, or at least to an open window. The hot liquid is allowed to cool naturally overnight (brewing is only done when the ambient temperature is agreeable), during which time the naturally occurring yeasts that live in the walls and on the rafters of the brewery inoculate the wort and begin the brewing process. The liquid is then transferred to barrels, where fermentation continues.

Brewing like that takes a lot of faith. Faith that the yeast will do what it's supposed to, and faith that nothing new and potentially off will be introduced—or even if it is, faith that it won't harm the beer too much. It's not uncommon for brewers who use this method to routinely spray down their walls with finished bottle-conditioned beer to help keep things in check.

As much as yeast has been cultivated, and as easy as it is for brewers to order what they need and have it delivered by FedEx, strides are being made every day to continue unlocking the secrets of this awesome little microbe. As a result, beers are being made with recently discovered yeast strains that introduce flavors never before imagined. In 2012, when Hurricane Sandy hit the New York metro area, Two Roads Brewing in Stratford, Connecticut, worked with local homebrewers in an attempt to catch yeast from the air during the storm. The pursuit was successful and resulted in a beer named Urban Funk.

Another innovative example of locally sourced yeast can be found in a beer created by the Lakefront Brewery in Milwaukee. I ran into Russ Klisch, the founder of the brewery, a few years back at the annual Craft Brewers Conference. Around us swirled conversations about hops, malt, and new equipment, but Klisch wanted to chat about yeast. "Everyone is currently concentrating on hops for flavor, but better and more diverse flavors can be had with different yeasts," he told me. Specifically, he was talking up a recently discovered strain, native to Wisconsin. Like many brewers, Klisch likes to keep things local. However, when it came to yeast he was stuck ordering from the catalogues, none of which were Wisconsin-specific. So he asked the manager of a local homebrew supply store who has a PhD in microbiology from Purdue University if he had an interest in finding a native Wisconsin yeast strain. The manager responded enthusiastically. The two began their search by looking at another beer ingredient: malt. Starting with a sample of Lakefront's barley, which had been grown and malted in the state, they crushed it and put it in a test tube with some water and nothing else. Yeast and other microbes started to grow, and Klisch's collaborator was able to keep it growing by feeding it more of the Wisconsin-grown barley. Klisch continued, "With his talents, he was able to separate the bacteria from the yeast and grow enough to do a homebrew batch." Eventually the whole process was scaled up for commercial use. The beer resembled a weiss, so they called the final result Wisconsin Weiss. It's been the brewery's summer seasonal offering for the last several years, with no sign that it's going away anytime soon.

Klisch, in the spirit of modern beer, thought it would be better to let everyone have access to the new strain rather than patent it, so the Wisconsin Weiss yeast strain is now available for purchase through Northern Brewer for homebrewers, or through Wyeast Laboratories for professional use. Says Klisch, "Yeast is the final frontier for the craft brewer."

There's practically an entire microscopic world of yeast strains to explore. Scientists have removed yeast samples from the bellies and wings of wasps to make beer. One brewery even took samples of yeast from women's genitals. Let's not dwell on the topic too much; suffice it to say that they crossed a line too far, backlash was swift, and even the most casual of polls revealed that there was nearly no market for such a thing. It was simply shock for the sake of shock.

Some brewers need only look in the mirror for inspiration. Take, for example, John Maier, brewmaster at Rogue Ales, in Oregon. In 2015 he created Beard Beer, an American wild ale made with Sterling hops, Munich, C15, and Pilsner malts—and yeast harvested from his own beard. Yes, you read that right. The yeast used to ferment the beer was plucked right out of the beard of the man who brewed it. Maier's beard dates back to 1978, according to Rogue. The label on the bottle prominently features a drawing of Maier's face (and beard) but doesn't offer many tasting notes. It encourages people to "try it, we think you'll be surprised."

In the name of journalism I accepted the challenge (mostly to save you all). I poured the Beard Beer into a tulip glass and, according to my notes, was "intrigued by its light bronze color and moderate white head. Orange blossom and honey aromas emerge from the glass. Makes me wonder if Maier uses a special shampoo." Upon sipping, I was met with "a full-bodied, slightly spicy, tangy ale with pronounced citrus and a bit of honey-like viscosity." I took a few more swallows, enjoying the flavors and trying not to think about the source of fermentation.

As the glass drains and the oddity fades, it becomes clear that great things can come from unexpected flavors. That Rogue experiment (along with others) could have very easily gone the other way and created an awful, undrinkable beer. Countless yeast experiments have wound up being poured down the drain, but that's par for the course when you're pushing the boundaries of an established product.

What I find most interesting is the chase to find new yeasts and to truly embrace spontaneous fermentation. You'll come across the descriptor "wild ale" quite often as you visit breweries and bars. Usually these are beers that have been fermented with *Brettanomyces,* a yeast strain that occurs naturally. If you're a wine drinker you might know it as a dirty word that can be a stab in a winemaker's heart. Thanks to the earthy, peppery, deep flavors it imparts (some use less-flattering terms, like "barnyard" or "horse blanket"), it has brought a new dimension to beer, capturing the attention of some brewers and drinkers. They ran toward it the way winemakers ran away from it (although much of the natural wine movement, where some "off" notes are open and acceptable, is thanks to beer), leaving some brewers to wonder if the notion of "wild" has lost its way.

I had a conversation with Brandon Jones, a long-time homebrewer and author of the blog *Embrace the Funk,* on the topic. He's now a professional brewer, blender, and keeper of the barrels at Yazoo Brewing in Nashville, and he wonders if the *Brettanomyces* available through a catalogue is truly "wild."

He and others are wrestling with this issue. They're trying to keep the tradition of spontaneity alive while also realizing the benefits of using cultivated strains. One answer has been to utilize mobile coolships, transporting wort onto farms and into fields where it can be inoculated in the true wild. Some brewers use this method to brew in state and national parks. These are ways both to find new flavors that hail from a specific place, and to tap back into our nomadic roots, a bit like our brewing ancestors. On a more commercial scale, a Nashville-based company, Bootleg Biology, is looking to single out and propagate a yeast strain from every zip code in the country, allowing not only professional brewers but also homebrewers to have a true "locally made" beer.

Everyone is interested in yeast these days. Unless, of course, you're drinking Bud Light. The best-selling beer in America touts four ingredients: water, malt, rice, and hops. No yeast. In a YouTube video that lasts fewer than fifteen seconds, the brewery reluctantly confirms that yeast is used in the brewery process, but is then filtered out, leaving just the other ingredients. I've wondered if the hard-working and classically trained brewers employed by that company cringed in the same way I did when they read that load of marketing bullshit. Whether it's still in there or not, every beer has flavors left behind by yeast.

We're drinking in the heyday of experimentation, from brewing with salty water to harvesting yeast from facial hair. As consumers, we should demand that brewers continue to innovate and improve while honoring the traditions and flavors embodied in the four main ingredients: water, malt, hops, and yeast. To do any less is to step farther away from what makes beer special in the first place.

THREE

IMPORTANT NOTES ON FLAVORS

THE FARTHER WE MOVE AWAY FROM DISCUSSING THE BASIC flavors of the four main ingredients, the more we verge into uncharted territory. While I'm all for exploration, and I appreciate the fact that breweries have introduced me to new kinds of fruits, blends of coffee, and other ingredients I might not have been exposed to otherwise, I also worry that each time a new flavored beer has a moment—say, a grapefruit IPA—and a flood of other breweries follow straight down that same rabbit hole, it gets harder, collectively, for them to climb out and come back to the fundamental recipes that make the foundation of beer.

I still prefer the classics, but I occasionally dip into flavored IPAs out of both professional duty and curiosity. I appreciate the natural grapefruit flavor that occurs in hop varieties like Azacca and Cascade, but in real life I don't enjoy grapefruit. I don't eat it with breakfast, I don't like it on salads, and the juice is something I'd rather pour down the drain than my throat. Still, I know what it tastes like and am able to appreciate how the flavor contributes to beer.

I'm about to say something important, so pay attention: there is a difference between a beer we don't like and one that's just not good.

QUALITY CONTROL IS AN ISSUE THAT EVERY BREWERY FROM BUDweiser to the neighborhood nano (generally defined as making two barrels or less at a time) must confront and maintain. Advancements in science have made testing equipment more readily available and more affordable as well. A lab setup can cost a few thousand dollars on the low end, but it's a necessary expense. Any brewer who has gone through professional training will have learned how to use even the most basic equipment that can examine yeast counts and other quantifiable aspects of the brewing process. Standardizing these across batches leads to better beer and helps brewers avoid unnecessary pitfalls. It's distressing to me, as a consumer, when I visit a brewery that's about to open and they are eager to show off their merchandise counter filled with hats and T-shirts, or a cool brewhouse toy like a hopback (a device that extracts essential oils from the hops to infuse a beer), or a spanking new forklift that's able to lift pallets of kegs—but they haven't put in a lab. This has been my experience more times than I can count.

The installation of a lab and the establishing of proper quality control methods should happen before a brewery opens, and certainly before any beer is released to market. Releasing for general consumption a beer that is not up to par is not only rude to consumers, but it hurts the overall industry. It's rare for any brewer to perfect a recipe on the first try, so a period of trial and error should go into each batch, which means taking a hard look at ingredients, processes, boil times, and packaging. Making sure a beer is dialed in on every level is paramount in today's market with so many breweries competing for the same dollars.

It's not that hard for brewers to meet these goals, although the process can be humbling. It's a good bet that the top beers on the market today began their journey by going through tasting panels, and not just with friends or family members who are prone to say overly positive things, but with industry professionals—people who've had taste training and are willing to give brutally honest opinions. Testing out recipes in advance of a formal release will ensure that a fine tasting beer makes it to a customer's glass time and again. There is rigorous sensory analysis that some breweries use, while some others will simply use a taproom to beta test a recipe. This ensures that the final beer has been tasted against other possible variations and is exactly what the brewery wants it to be.

Completing a training program like the Cicerone Certification Program is a good idea for brewery owners, brewers, and other staff. The curriculum is sensory heavy, taking a deep dive into individual beer styles by examining the brewing process, ingredients, and history behind them. It even gets into the mechanics of cleaning draft systems. Kits are available—for both large-scale brewing concerns and homebrewers—to help brewers identify hop and malt varieties, as well as off flavors like diacetyl (see below). Education and testing should never stop. Quality control is nonnegotiable, and constant monitoring is mandatory.

It is also important to step away from the ingredient component and look at delivery methods. Even large breweries with solid quality control protocols are always looking to evolve. Sierra Nevada Brewing Company, for example, found that the taste of their beers improved after the brewery switched from twist-off caps to the fitted variety.

Perhaps nothing is more important or affordable than experience. Many of the breweries that made it through the 1990s shakeout were those whose owners or brewing staff had extensive professional or homebrewing experience—people who knew how the beer should taste and weren't afraid to dump a batch if necessary (even considering the financial loss) because they knew that an ill-flavored brew wouldn't fly.

THAT SAID, WHAT WE CONSIDER "BAD" BEER IS IN FLUX THESE DAYS. Hang around a serious beer bar or a beer geek long enough (or not that long, usually), and you'll hear the phrase *off flavor* being muttered. But just what is an off flavor when it comes to beer, and how does it get there?

Off flavors can occur due to poor sanitation of equipment, poor brewing techniques, miscues during fermentation, and a host of other reasons. These errors can result in chemical compounds or chemical reactions that shouldn't be in the beer, for example, acetaldehyde, chlorophenol, diacetyl, dimethyl sulfide (DMS), and oxidation. No matter the cause, once any of these substances is present in the beer, they can be difficult (or impossible) to erase. They can impart characteristics that are medicinal, metallic, skunky, soapy, or sulfurous. Those last five words describe aromas or flavors that are easily identifiable, and although they are appropriate in certain contexts, when they appear in beer it can be to the great frustration of the brewer. Just as it's important to use descriptive language for the main ingredients, the same is true when it comes to these chemical compounds that sometimes appear in our beer. Acetaldehyde in beer tastes or smells like green apples or fresh-cut pumpkin. Chlorophenol presents as Band-Aid. Diacetyl and DMS are the two we're most likely to encounter as drinkers; respectively, they smell like movie-theater popcorn and creamed corn.

But wait, you say. I like butter on my popcorn when watching the latest superhero movie. And creamed corn was a delightful vegetable when I first tried it as a kid. Why should we not want these flavors in our beer? Why is it a pleasure to bite into a Granny Smith apple, but not have the same taste in our glass?

It basically comes down to the styles of beer that have been established over the centuries. We have collectively decided, over time, what an IPA generally tastes like. While we might like melted butter on our lobster, when it shows up in our stout it detracts from the beer's core ingredients and changes the overall flavor profile. Same with green apple in our lagers, or creamed corn in our pale ale. The emergence of these flavors also speaks to the skill with which the beer was made. A good brewer knows how to prevent these flavors. There have never been more troubleshooting resources available to a brewer than there are today. Just about any conceivable problem already has an answer, and it's as close as doing an internet search on a brewer's forum.

Still, these problems persist. One of the great privileges and perks of my job is visiting breweries around the world. It allows me to meet brewers, drink beer, and see the innovations, ingredients, and trends that help shape my general news coverage of the industry. On a trip in mid-2016 to Kentucky, I ordered a flight of beers from a brewery that had opened its doors to the public twenty-four hours earlier. The room was buzzing with a happy energy despite gloomy weather outside. Staff members were busily attending to visitors while still working out answers to everyday questions, and a whiff of fresh paint lingered in the air.

The beer itself was fine. The lineup was what you'd expect from a US brewery these days: a few IPAs, a Scotch ale, an amber, a wheat.

My notes from the day express appreciation for some, a few flaws in others (like DMS in a pale ale), and some technical issues (like low carbonation), that would likely be addressed down the line. At one point the brewer walked up, looking exactly like a brewer who had just run a marathon to the finish line of opening a brewery. Without prompting, one of the first things he did was apologize for one of the beers on the table. It's not how he'd hoped the recipe would turn out; he knew there was an off flavor; he'd work to get it better the next time. That admission represents the good and bad about professional brewing. Good because he has the desire to do better, to push his talents, and to offer customers a stronger product. Bad because you only have one chance to make a positive first impression, and to knowingly put subpar beer into the glasses of paying customers is an affront to the whole industry.

SOME OF THE OTHER OFF FLAVORS, CAUSED BY PROBLEMS LIKE light exposure (termed "lightstruck") or oxidation—which show up as skunk and wet cardboard respectively—in fresh beer should be avoided. But these flavors can be acceptable in, say, a saison, in which a little lightstruck aroma is allowed, or a barleywine that has aged for a decade and has taken on a cardboard flavor that's also a bit like sherry.

Then there are brewers who want to embrace certain off flavors. Why can't we brew a lobster-infused stout (an offshoot of the more traditional oyster stout) that has a diacetyl note? Why not make a casserole beer with all the trimmings? The answer is: you can, and there's likely a niche of customers who will flock to it and might even enjoy it. It's just unlikely to win many awards from the Beer Judge Certification Program. These are novelty beers, fun to drink now and again, usually a few ounces at a time. When it comes to the every-day-drinking beers, the ones we order by the pint at the bar, it's up to us—the drinker, super beer geek or not—to know the basics of off flavors. The more often we allow inferior beers to be made and served without alerting the brewer—who may not know about the bad batch, or may simply think the beer passes muster—the more harm it can do to other breweries and to beer in general. So if you smell a beer that has aromas of green apple, mildew, or fake butter, recognize that it's likely a flaw. A little politeness goes a long way, and if you're served a beer with one of those traits, find a way to personally bring it to the brewer's attention, or at least the bartender's.

Hopefully they will be receptive, but be prepared for the other version of that conversation. I was at a brewery in southern Florida a few years ago, sitting at the bar waiting to meet up with an award-winning homebrewer who lived in the area. I ordered a pint of brown ale and was disappointed to be met with sticky, chemical fake-butter aromas. I quietly got the bartender's attention and mentioned the flaw and asked for a different beer. He took it back, but not before explaining to me, in a condescending manner, that this particular brewery was helmed by an accomplished and respected brewer and that my assessment—whoever I was—was dead wrong. He'd get me another beer, but he was going to charge me for the first one.

Mind you, I hadn't revealed my identity or profession (because, let's be honest, no one likes the "Don't you know who I am?" guy). My contact, the homebrewer, showed up, and, as he was a regular, said hi to the bartender. The bartender then pointed a finger at me and said something like, "Your pal here seems to think he's some kind of expert and had unkind words for our beer."

I stayed silent as the shocked homebrewer mentioned my

credentials, and the bartender's face dropped. The brewer himself arrived at our station a few minutes later with an apology and an acknowledgment that this particular batch of brown didn't come out right. Then he made a big show of taking it out of rotation on the tap lines.

I didn't aim to be a jerk. I just don't like paying for bad beer. Neither should you. As more drinkers become educated about clearly unintentional off flavors (which hopefully will happen less often as brewers get their practices dialed in), and if we speak up when we notice these problems, over time the beers will get better. That's good for everyone. In the meantime, if you get a bad pint it's not unreasonable to ask for your money back.

This is an important point that many of us might forget now and again. We customers are the ones with the money and therefore the power. If a brewery doesn't offer you money back on a subpar or just plain poor pint, there's a good chance you won't be going back there again. Quality is key. Without it a brewery doesn't have a good name to stand on and will soon find itself out of business.

As mentioned above, however, it's equally important not to confuse proper quality with our personal perception of taste. Because taste is not always what we think. Take the example of watermelon: fresh-cut and juicy, a vibrant pink on the inside, garden-green on the rind. A staple of summer cookouts and a wonderful pool-party snack. What does it taste like? Can you pick it out in your brain? Now think of a watermelon-flavored Jolly Rancher or other hard candy. Is that easier? I thought so. The truth is that even the juiciest watermelon packs some flavor, but not a ton. Candy makers are aiming for strong flavors, so they take the "better living through chemicals" approach: they concoct a group of flavor compounds that draw inspiration from the original fruit, then add extra sugars, until they've come up with the all-too-familiar candy version. Watermelon hard candy is light years apart from an actual watermelon, but everyone will still say, "It's watermelon." This is also the case with raspberry: the candy form is usually blue in color (to help us visually differentiate it from other red berries), yet we don't question it. The same is true for banana. The candy version is a far cry from the fresh jungle fruit, but we, as consumers, have come to accept both as equals in terms of taste.

Chemical flavoring is having a big impact in beer these days. An industry built on tradition, quality, and pure ingredients is beginning to regard chemical flavorings as acceptable. Brewers of watermelon wheat beers might throw in some fruit as a token gesture before adding gallons of flavored syrup to the fermenters. Grapefruit-flavored IPAs continue to be popular, thanks to the citrus flavor that naturally occurs in some hop varieties. But lest you think large brewers are sitting around cutting up grapefruit for the mash, the reality is that most are adding grapefruit powder, and in some cases it's the same kind that goes into Flintstones-brand vitamin tablets. A decade ago you'd be hard pressed to find flavoring companies at the annual Craft Brewers Conference, where brewers gather to purchase equipment and attend educational seminars. Today, dozens of these companies show up at the event to hawk their flavors—from "birthday cake" to "black cherry" to "cinnamon challenge."

In some ways it's disheartening to see brewers flock to these companies, to know that centuries of brewing progress are being mucked up by the same flavors that made vodka silly and made coffee taste like donuts or salted caramels. It's also a sign of the times. If a brewer can make money, keep the lights on, and grow a business by selling an Almond Joy–flavored blonde ale, what's the harm? They're offering a product to a consumer who might prefer a familiar flavor from the candy aisle over an aggressively hoppy double IPA.

That's not to say that long-hardened beer fans don't enjoy some of these specialty flavors. Brewers, many of whom have been in the industry for a long time, have taken to experimenting with beer flavors and creating recipes that are supposed to taste like something else. I've seen beers flavored to taste like German chocolate cake, key lime pie, panzanella (bread and tomato) salad, pastrami sandwich, even cocktails like an old-fashioned or a bloody Mary. Curiosity eventually gets the better of us all, and tasting these beers not only is a fun sensory experience, but can bring back dormant memories of favorite foods. I once had a beer made with caramel malts and almond extract that reminded me of the cookies served by our local Chinese restaurant after dinner. It had been years since I'd eaten that dessert, and the taste of the beer took me down an unexpected memory lane of family gatherings.

The one key, brewers say, is that no matter the ingredients used, flavorings added, or homage paid, the final result should still taste like beer. To create something otherwise takes us too far afield from the true spirit of brewing, to the point where you're just making a flavored malt beverage (FMB), something like a Twisted Tea or hard lemonade. Love them or leave them, those products should never be considered beer, because they aren't.

The other key thing to consider is that certain flavors, no matter how delightful to some, won't work for others. I, for one, have an aversion to Mosaic hops. I know how it tastes to me, and I'd rather not taste cat box when drinking. I'm jealous of folks who get to enjoy beers made with this hop (mostly because there are so many of them out there these days), and who pick up on the pineapple and tropical fruit flavors. The hop is so popular, especially with newer IPA-focused breweries, that I often have to completely steer away from hoppy styles when visiting certain breweries for fun, or I have to taste until I find the least offensive (to me) beer in the batch.

There are other polarizing beer flavors that some people might love, while others certainly don't. Beginning in late July or early August, pumpkin beers begin arriving on shelves. I know that some people love the combination of spices that make up what we think of as "pumpkin": nutmeg, cinnamon, clove, allspice. They love the warm, inviting aroma. It conjures up thoughts of Thanksgiving dessert, of colder months spent indoors under warm blankets. But did you know that beer is largely responsible for the proliferation of these flavors outside of pie—and for them ending up in your latte? In 1985, Bill Owens, the man behind California's Buffalo Bill's Brewery, released a pumpkin ale. Made back then with actual pumpkins (you don't actually think fresh gourd is going into those present-day summer releases, do you?) and spices, it was a huge hit and is still made today. Other brewers followed suit, including Dogfish Head, which released one in 1995. (More recently, for a while it seemed like every brewery in America made at least one pumpkin beer, and others, like Elysian Brewing, released at least a dozen annually.) Starbucks, which gets so much of the credit for this flavor combination in anything other than dessert, didn't release a Pumpkin Spice Latte until 2003.

I'm a fan of baked goods. But I do not have a taste for those spices when they are combined. Moreover, so many of the pumpkin beers released each year are anything but subtle, resembling full bags of spice clumsily dropped into the mash. I've said on the record—and I stand by my statement—that every pumpkin beer should be rounded up, put in a giant pile, and set on fire.

When I say this, there's always someone who asks me to try a new one. And I oblige because it's the right thing to do. To date, the only pumpkin beers I've found that I can mostly tolerate are the ones that come out in late October or early November and that use real pumpkin, placing that ingredient in the starring role as opposed to the spices. Sadly, such beers are few and far between. They are based on subtler recipes that won't score well with the Pumpkin Spice Latte fanatics who not only celebrate the coffee but buy Oreos and lip balm that's been subjected to the same treatment.

I will never begrudge people who truly love something, even when I don't. That's a good life motto, and it's also a great way to approach beer. Because not every beer style is going to be appreciated by everyone. Long ago I realized that there are certain beer styles, such as Belgian tripels and quads, that I just don't prefer. Yeast esters that are fruity and perfumey, as well as those that produce a higher ABV (alcohol by volume, expressed in terms of a percentage) and robust carbonation, simply don't suit my palate. Professionally, I learned how they were supposed to taste and the processes used to make them, but if I had the opportunity to judge a beer contest where they were offered, I'd beg off unless there was no alternative—in which case, I'd give it my honest shot to make an assessment uninfluenced by my personal bias. Doing so required more concentration than judging beers I enjoyed, and I'd always be honest with the other judges about where I was coming from. In many ways, being forced to think about and try to appreciate these beers made me a more open-minded drinker and exposed me to new flavor profiles. Similarly, I would encourage you to try more than once a flavor you don't like. Our palates are constantly changing, and many wonderful beer flavors are indeed acquired tastes.

Sometimes, when it comes to identifying off flavors, you will have to trust others around you. Earlier I mentioned diacetyl, the chemical compound that tastes like movie-theater popcorn. Well, I don't have a good receptor for it. Some folks are very sensitive to its presence and can pick up the smallest amount with the faintest of whiffs. Others notice it when a beer is put through the normal tasting process. For me, unless a beer is teeming with the stuff—and I mean to the extent that the sensitive folks can smell it from across the room—I won't notice it. I've tried, but my receptors just aren't wired that way. But they are for oxidation and for acetaldehyde, the green-apple flavor. I can pick up those aromas quite quickly. It's important to remember that when it comes to tastes we're all wired similarly, but we also have individual quirks. Time, a lot of tasting, and a willingness to learn and evaluate help each of us find what we like, what we don't, and why.

This topic leads to something I find particularly frustrating about online reviews, especially the rating sites. It's one thing to give a low score to a beer that clearly tastes stale, or one that is labeled a hoppy IPA but has no hop aromas or flavor. It's another thing—and if you look closely you'll see this happening all too often—to rate a well-made Kölsch with a low score because you only like stouts. Or to judge a classic pale ale against a more modern New England style. The rule I've always observed when tasting and evaluating beer is that each should be its own island. When I was writing reviews for a magazine, the idea was never to compare one beer against the other, but to discuss each based on its own merits or flaws. This is really how online ratings from consumers should work, rather than allowing personal biases to cloud one's judgment regarding an otherwise great beer.

Spend some time in certain beer circles these days and you'll hear people using words like *bugs, funk, brett* (short for *Brettanomyces*), *puckering,* and *tart*. They are talking about wild beers, a catch-all term (which often incorporates spontaneously fermented beers) that is getting quite a bit of attention. Because of its growing popularity, many speculate that wild beer could be the next IPA in terms of generating a cult-like following. But because they are so largely misunderstood, so extreme, or just so different, what would be a flaw in another beer is celebrated in the sour. In the same way that hoppy beers were so polarizing in the beginning of the modern beer movement, the equivalent is true for sours today. But in the same way that every brewery quickly came to embrace hops, now nearly every brewery offers at least one sour. Some breweries have such extensive sour-beer programs that they split their lists between "wild" and "clean." With each new entrant to this arena, there's a chance for experimentation and increased understanding, and a way to dial in what makes a sour a sour.

In modern beer terms, *sour* typically means "acidic." The primary difference between sours and the more common ales and lagers is, simply enough, the pH, or acid content. Most beers fall in the pH range of the low 4s. Sour beers are much more acidic, with a pH that may be as low as the mid 3s. While that might not sound like much of a difference, every full point on the pH scale indicates a ten-fold difference in acidity.

Sours start off like every other beer, and they contain the same ingredients—water, malt, hops, and yeast. What gives a beer its sour disposition is the final addition of microbes. Most sours start out using the yeast strain *Brettanomyces,* which inspires descriptions like "barnyard," or, more specifically, "horse blanket." Not the most appetizing imagery, but not inaccurate. In addition to these earthy qualities, *Brettanomyces* also produces notes of spicy pepper and even tropical fruit, both of which lend real character to beers. After the yeast, it's the addition of *Lactobacillus* or *Pediococcus* bacteria to the beer that increases acidity in the final product, qualifying it as a "sour beer." The bacteria are typically inoculated during the aging process, which generally takes place in barrels over a generous amount of time. Interestingly, certain strains of wild yeast, recently discovered, produce *Lactobacillus* bacteria as part of the fermentation process. If these yeasts were put into wide release, the result could be a game changer for the sour-beer style.

"Generally speaking, making a beer is about a thirty-five-day process," Patrick Rue, founder of The Bruery in California, told me a few years back. For at least one of his sour offerings, the beer ferments in a puncheon for two months and is then transferred to smaller barrels and aged sixteen months before being bottled.

Among the most common styles of sour beers are lambics (both with and without fruit added), Berliner weisse (which has a yogurt-like characteristic), Flanders red (stone-fruit aromas and flavors), and oud bruin (balsamic notes often present). Most are actually low on bitterness, with the acidity balancing their malty sweetness. Often they finish very dry. It is not uncommon for brewers to blend older batches with younger batches to create a more palatable balance. This happens with lambics as well as other styles, including Flanders red and oud bruin.

While the sour is not a style that the casual beer drinker is likely to start with, those with experienced wine palates often appreciate the great charm of sours, especially wine drinkers who gravitate toward traditionally sweet varieties like riesling or muscat. Perhaps sour beer's appeal to these wine lovers arises from the fruitiness exhibited by the yeast, or the dry finish, its acidic nature, or its ability to age well over time, or how it complements a variety of cheeses.

Most breweries have a sour program, but that doesn't mean all their sour beers are worth tasting. Many are still finding their way. Established breweries and those with strong brewing talent who know how to manage the bugs—including, in America, Allagash

Brewing (Maine), Russian River Brewing Company (California), Jester King (Texas), De Garde Brewing (Oregon), Jolly Pumpkin Artisan Ales (Michigan), Hill Farmstead (Vermont), Black Project (Colorado), Yazoo Brewing Company (Tennessee)—are producing rare, sought-after sour beers. Rare because of the aging time they require, which means that brewers often cannot release large quantities of these specialty beers to the general marketplace, especially if they need production space for other beers with quicker turnaround times, like stouts, lagers, and IPAs. Some breweries have recently begun dedicating brewing space or even entire breweries to the production of sour beer.

"It's not a great, appealing name. 'Sours' is not a particularly glamorous word," Michael Tonsmeire, author of *American Sour Beers: Innovative Techniques for Mixed Fermentations,* said to me. Still, "there are some that are very approachable, if you like tart and refreshing." Similar to how we should use specific flavor descriptors for IPA rather than just "hoppy" or "bitter," remember Michael's observation as you encounter more of these "sour" beers. While sipping one, see if you can identify the variety of flavors and aromas it offers. It will often take several tries (or longer) to start to pick out nuance.

ULTIMATELY, WHEN IT COMES TO BEER, DRINK WHAT YOU LIKE. Know yourself and your palate. If you don't like chocolate desserts, maybe stay away from the chocolate porter. If you favor Cheerios in the morning, check out a helles lager. If you like the sweet nature of graham crackers, give a Berliner weisse a try. No matter what, don't stop exploring or trying new beers. There are endless flavors available, and although all of them won't appeal to you, aim to avoid falling into a rut of sameness. Being just a little bit adventurous can lead to a new discovery and a new passion. Each new beer tried and evaluated makes you a stronger drinker, allowing you to celebrate a brewing triumph or spot a flaw. Your explorations in the beer world not only keep traditions alive but also encourage experimentation. And that's what propels beer forward.

FOUR

HOW WE DRINK BEER

LIKE AN INVISIBLE FENCE ERECTED ACROSS RESTAURANT AND dining room tables, bar tops, store aisles, and at social gatherings, there has long been a division in the beverage world. On one side were the beer people. Wine folks populated the other. It was uncommon, almost unheard of, to move from one side to the other. Then along came the millennials, and with them, the concept of cross-drinking. In the past, one person would likely drink beer (or wine) all night long, never switching from their original choice, but cross-drinkers are more flexible: they might start an evening with a cocktail, then move to a beer, hop over to a wine with dinner, and finish the night with, perhaps, a hard cider. The alcohol companies bemoan this practice because it leads to a lack of brand loyalty, and potentially lost sales. In this age of abundant choice, however, I think it's great. And while I'm not myself a millennial, I've wholeheartedly embraced the practice.

There's a lot to learn from trying new wines, spirits, or cocktails. Cross-drinking's true benefit is that it helped normalize beer in a society that still hadn't quite accepted it. Whereas for a long time beer was seen as the lesser beverage, it's now on an equal playing field. This is largely due to the increase in smaller breweries, which are producing higher-quality beers than those of the past. But it's also due to *how* and *where* these beers are being served. Beer isn't wine and shouldn't try to be. Still, brewers and drinkers can learn a few things from vino's long social dominance and from the fine attention to detail observed by oenophiles. These are the qualities that gave wine its regal and sophisticated reputation, and they start before you even take a first sip.

The decision to order a beer, and which beer to order, is formulated deep in our brains. Spend enough time at a bar or brewery and you'll start to develop cravings for a certain style, a certain brand, or a beer that you know, deep down, will soothe your soul, lighten your mood, or even help you brood.

Beer makers devote a lot of energy outside the beer-making process trying to find ways to lure you in. They use traditional advertising in magazines, yes, but they also reach you through tangible things that have nothing to do with the liquid itself. I'm thinking of everything from the logos on coasters, to the shape and billboard-like advertisement of tap handles behind the bar, to even the glass you hold in your hand, if it's designed with a particular beer in mind.

Each time we order a beer, we think we're making an intellectual choice, but it's actually very largely hidden from us. It can be an involved one that is the culmination of a career spent drinking beer, and is based on your mood, the flavor you want, or the type of refreshment you're looking for (a lager on a hot summer afternoon, or a barleywine next to a fire while a winter storm rages outside). It can arise from the curiosity of trying something new that you've heard or read about, or from simply knowing that you want a certain style.

When we order a beer—be it by specific name or general style—something pushes us toward that choice. It might be a beer that is familiar, or one we've seen advertised, or one with an inviting description on the label, or with a good menu description. As we become more aware of our personal tastes and preferences, we make decisions based on individual ingredients, or we find a beer that fits a mood or meal. From there the next logical step is to encourage ourselves to step out of our comfort zones, go back to styles we once ignored or dismissed and see if we're ready to be enchanted. Here we can be a leader, encouraging not only ourselves but others to explore new styles, beers with different looks and odd ingredients. When we can be genuinely excited about the beer we are ordering, no matter what style or who makes it, the more pleasurable overall the drinking experience becomes.

As the writer Randy Mosher once reminded me, "Brand preference can be based on personal history, peer pressure, social ambitions. perceived coolness, advertising, familiarity, packaging, social media, and a host of forces that are anything but rational. This has all been well-studied and as a drinker, it's good to be aware of this fact, because most of us shut ourselves off from new experiences and fail to appreciate the ones we have because of our accumulated biases and preconceptions." Remember from the Introduction how I mentioned the "mind palace," introduced in the TV series *Sherlock*? It's where Holmes mentally retreats to gain access to information stored in his brain. I said that I would ask you to create a "mind pub," a perfect-to-you drinking spot. Now, go to your mind pub and order a beer. Maybe it's sliding toward you across the bar from your local bartender, handed over on a friend's front porch, or sitting on your own kitchen counter after you've poured it yourself. No matter the setting, there should be a fun moment of anticipation before you pick up a freshly poured glass of beer. Picture it now. Can you see the carbonation rising and a collar of foam? What color is that liquid, and is it clear or hazy? This beer you're picturing: is your gustatory memory (the fancy term for flavor recognition) already revealing what you know it tastes like?

Now, tell me about the glass. Not the contents, but the actual vessel. Proper glassware is just as important as the liquid it holds, but it's often overlooked, especially here in the United States, where it's considered an afterthought.

Beer arrives to us in one of three main ways: on draft via kegs, in a bottle, or in a can. In certain situations it's fine to drink directly from the bottle or can, like at sporting events, or in social settings where glassware isn't available or practical. Some beers are even designed to taste better from a can. In the case of the New England IPA, a hazy hop-forward concoction that resembles pulped juice—meaning it clearly contains a bit of yeast slurry, hop residue, or even flour to help with mouthfeel—it's better to drink it directly from a can (this style is almost universally canned) because aesthetically it's just not great to look at. A pioneer of the style, the Alchemist, in Stowe, Vermont, has written instructions to "drink from the can" across the tops of its Heady Topper cans.

But whenever feasible, it's best to use a glass. And whenever possible, that glass should be the right one. America has an obsession with the shaker pint glass. That's the cylinder-shaped 16-ounce (usually) workhorse, ubiquitous wherever draft beer is sold. It's familiar in the way we all just accept bland, always disappointing airline food. If it doesn't leak, it's good enough for the job.

That doesn't mean it's good enough for the beer it holds.

Bar owners like shaker pints because they're stackable behind the bar, where space is at a premium. They're durable enough to survive short falls, can be used for soda and water as well as beer, and were designed to fit the metallic Boston shaker, making the pint glass just right for whipping up cocktails.

There are downsides, however, to this style of glassware, and ultimately it stunts the beer. When a brewer releases a beer into the wild, they hope it will arrive at your taste buds in a top-quality condition. But a lot of factors can go wrong between creation and delivery, from less than ideal temperature variations, to too much time passing between when it's made and consumed. Often it's the final step—the pour into the glass—that goes awry.

When it comes to tasting beer, aroma is paramount to the full experience, but the olfactory sense doesn't always get the chance it deserves. In America, when we order a pint we want 16 ounces because that's what we're accustomed to. That's what we've paid for, and you can bet there'd be complaints galore if space were left at the top of a glass to properly release aromas. No, we want our shakers filled to the brim, and a little sloshing over is acceptable because it proves we weren't shortchanged.

Now, try to get a whiff of the aroma coming off that spilling-over beer. Get your nose close and . . . oops! Now you have a wet schnoz.

While I'd like to think that every beer is created equal in quality and is deserving of the proper olfactory rituals, we know that's simply not true. A filled-to-the-top pint of Bud Light or Heineken doesn't need to be fussed over or sniffed. Chances are you've had it a hundred times before, you know what you're getting, and you just want to drink it, not necessarily experience it. Yet for the more aroma-forward beers—like IPA, saison, barleywine, or any other style with ingredients that add a pop of depth and flavor—space at the top of the glass matters.

For the amount that the wine industry charges for a bottle, you can bet they've put time and attention into making sure the glass you use suits the liquid. Fine wine is seen as a luxury, and the allure would be lost if you just reached for a red Solo cup. Similarly, to fully appreciate the beer—appearance, aroma, and taste—you need the right glass. In recent years a number of glassware companies behind high-end wine and spirit glasses have stepped into the beer arena to add beer glasses to their portfolios. Usually they're made of higher-quality glass, are thinner (to help regulate temperature), and a good number are stemmed, just like wine glasses.

Beer has taken a page from vino's book, and there are now a number of glasses designed for specific beer styles. The one that made the biggest splash a few years ago was an IPA-specific glass, manufactured by the German company Spiegelau with input from both Sierra Nevada Brewing Company and Dogfish Head Craft-Brewed Ales. Since then Spiegelau has partnered with other breweries to create stout, barrel-aged beer, wheat, and lager glasses. Although it might seem like an extravagance to buy specific glassware for your beer, as a person who's done blind taste tests and run temperature trials, I'm convinced that a quality glass leads to a better beer-drinking experience. Plus, I do enjoy the aesthetics of the glassware itself. This isn't a commercial for Spiegelau, though. Other companies have gotten into the beer act, like Luigi Bormioli and Libbey (the largest maker of that ubiquitous shaker pint).

It's more common to find style-specific glassware at breweries than bars. If you get your hands on beer-specific glassware—and let's just focus on the IPA glass here—the first thing you're likely to notice is how thin the glass itself is. This is by design; it helps keep your beer cooler and the aromas fresher. I know that for the science-minded readers that statement makes sense, but if you're like me and are surprised by it, read on.

The glass of a shaker pint is about one centimeter thick. The Spiegelau glass is half that, and that thinness makes all the difference. Think back to the last time you were at a bar and drinking from a shaker pint. By the time you were halfway done, the glass was warm to the touch. Not only does the thicker glass absorb your body heat; it also pulls the cold away from the beer. Both phenomena cause the beer to warm up faster, ultimately dampening the aromas and flavors. Thinner glass draws less of the cold temperature away from the beer itself, keeping it cooler longer, and thus closer to the brewer's intention.

The thin glass also allows us to see the beer more clearly. In a side-by-side comparison between the shaker pint and a thinner IPA glass, you would notice that the specially designed beer glass gives you a better look at carbonation, color, and other fine visual details.

This is like making the switch from the old tubed televisions to high definition: once you've noticed the difference, it's hard to go back. It's not unlike drinking wine from a fine stemmed glass versus a jelly jar.

Beer-specific glassware affords a few other perks that enhance the drinking experience. The IPA glass has a ribbed base that agitates the beer as you lift the glass, releasing more hop aroma. The bowl shape on top of the base gathers these aromas and releases them to the olfactory receptors, giving the drinker a deeper experience. The stronger olfactory experience is boosted by the smaller opening at the top (two and three-quarters inches versus three and a half inches for the pint glass). When you put the IPA glass to your lips and tilt it back, the opposite rim touches the top of your nose, creating a contained area. A pint glass, by comparison, hits your glabella, or that space between your eyebrows. The pint glass's wider mouth means there's more room for the aromas to escape. And if you tilt your head back too far, it's easier to spill. Yes, all this does sound fussy, I'll admit, but in my opinion it's the right way to treat a beverage that's been made with a lot of care and attention. Drinking from an appropriate glass enhances a personal experience on many sensory levels, and elevates what you're drinking to something just a little more special.

Preferences for certain glassware, like other tastes, is personal. When you're at home you might have a preferred glass that doesn't do the beer any favors, but reminds you of someplace or someone special. Perhaps it's a pint glass that you nicked from a bar in college, or a stein that belonged to your grandfather, or an oddly shaped glass handed out at a beer festival you attended long ago. If that glass is what brings you the most enjoyment from your beer, by all means use it.

In general, though, I feel that if you're spending your hard-earned money on a quality beer, it deserves a good glass. The new glasses might not be the workhorses that we're used to, but they are more durable than you'd think. Although all glassware is susceptible to chips, cracks, and full-on shattering, the thinner glass can handle your at-home drinking sessions and celebratory clinking with ease. It should also go without saying, but if you're served a beer in a chipped or cracked glass, immediately send it back. Our bodies weren't made to process glass.

At home, I generally use a tulip glass that's been properly washed and rinsed. The shape of the stemmed glass is much like the bulb of its namesake flower, bulging out in the middle from a tapered top and cupped bottom. Our cupboard is still filled with pint glasses (my wife and I even had some made up as our wedding favors), but often we use those for seltzer. The tulip glasses I regularly use can hold up to 12 ounces of liquid, but they have an indicator at the 10-ounce (0.3-liter) mark, where I usually stop pouring. I can always add more beer to my glass from the bottle or can later, but by hitting this mark I'm able to ensure that there's a good amount of space between the top of the foam and the lip of the glass. It's functional, it looks good, and it works for a variety of styles from Vienna lagers to imperial stouts. Also, it's affordable. Overall, it makes for a better drinking experience.

THE SHAKER PINT IS MORE THAN JUST A VEHICLE FOR BEER—IT'S also a vehicle for marketing. Think of the shaker pint glass as a billboard. We see them along the side of the road each day: same basic shape, usually up high, but the message is different. Even with the rise of high end beer bars, you're still hard pressed to find a drinking establishment in the country that doesn't use shaker pints. Breweries know this, and even if they are particular about how their beer is served and wouldn't use the shaker pint in their own taprooms, they aren't shy about using the real estate on the side of the glass for their logo. Next time you're out, take a look around and you'll see all manner of logos (and not just for beers and breweries but for sports franchises, restaurants, charity events, and just about anything else you can imagine a company would want to get in front of a captive audience). Sometimes brewery-branded shaker pints are part of a promotion (when permitted by state law): if you order a pint the glass is yours to keep.

Even though it goes against the seventh commandment, those logo glasses often find their way into purses, coat pockets, and briefcases, usually after a few rounds, when inhibitions are down. While the practice is officially frowned upon, most brewers generally don't get too upset about the theft. The idea is subliminal. If a shaker pint glass stamped with, let's say, "Founders Brewing Company" finds its way from the bar to your kitchen and becomes part of your regular rotation for everything from orange juice and milk in the morning to iced tea or soda for dinner, you'll see that logo over and over again.

So next time you're in a bar or restaurant and see a Founder's tap handle, you might gravitate toward it. Clever, right?

Let me be clear that I'm not advocating theft—I've found that by simply asking a bartender you can usually walk out with the glass guilt-free. But I'll admit that I'm not without sin in this regard. My first postcollege apartment was well-stocked with pilfered glasses. My sincere apologies to all the affected bars.

Travel throughout parts of Europe, and although you'll still see logo glassware, it's harder to find the shaker pint. I'll single out Belgium and Germany, countries where breweries are more likely to have their own glassware—a shape unique to that brewery—as well as taverns that actually use the branded glassware. These international breweries with their own signature glasses benefited from a long and uninterrupted history of beer-making (unlike us in America), and realized early on the benefits of signature glassware to increase visibility and sell more beer.

Orval is a classic example. Orval is the namesake beer of the Brasserie d'Orval, an abbey brewery in Belgium's Gaume region, on the border with France. They produce a Belgian pale ale, bright amber in color with a persistent fluffy white head of foam. Thanks to *Brettanomyces* yeast, the taste is both earthy and spicy. An aged hop content offers hay and slight citrus. It finishes dry. The beer has a shelf life of up to five years, during which, if cellared, it continues to age in a subtle way. After that, it falls off dramatically. Orval is a universally celebrated beer, thanks in no small part to Michael Jackson (the beer writer), who in the 1970s and 1980s introduced the then relatively small and obscure beer to his wide international audience. The beer is best sipped from its signature glass, and, at least the first time, should be enjoyed in a location where each sip can be contemplated. It has inspired generations, and it regularly places at the top of other brewers' lists of personal favorites because of its intriguing nature. If you've never had an Orval, stop reading right here, go get a bottle, and come back. Actually, same goes if you've had it once or a thousand times. The rest of this discussion can wait.

The Orval glass is a uniquely shaped vessel that has been produced by the brewery for years. A wide conical-shaped chalice with a silver-painted rim sits atop a solid column of glass and a base. "ORVAL," in silver-outlined black letters, is written in a size that's easily readable from twenty-five feet away. Some of the chalices display the brewery's logo—an upturned fish holding a ring in its mouth against a blue diamond—which is also instantly recognizable.

Beer drinkers like to argue about whether the Orval chalice actually enhances aroma, flavor, or appearance as well as some of the modern glassware does. Breweries in Europe reject those skeptics out of hand. Either way, using a glass intended for a specific beer is likely more about a feeling, an intangible pleasure that makes us feel closer to the contents of the chalice.

Specialty glassware isn't limited to specific breweries; it is also created to match certain styles. In Cologne, Germany, the local beer is Kölsch, and at bars around the city it's served in a stange, a tube-like glass that holds 200 milliliters, or just under 7 ounces. Order one, and servers will keep bringing you full glasses as your current one empties until you tell them *bitte nicht mehr*. Weizen beers, like hefeweizen, often come in a vase-like glass that can hold 23 ounces. (You might also see it served in a boot-shaped glass. This is more for the visual effect than anything else, but it makes for a good Instagram shot.) The weizen vase glass is also usually garnished with a lemon wedge. When you use a specialty glass, not only does what you're drinking get signaled to other patrons, thanks to the shape of your glass, it also creates a communal atmosphere in a way that the shaker pint just can't, because it signifies that you're all drinking the same thing.

One of the lasting legacies of post-Prohibition brewery consolidation in the United States and the one-size-fits-all beer is also the one-size-fits-all glass. However, the rise of the modern beer movement and of smaller breweries has led to the arrival of small-batch glassware. Usually these glasses come from artisan producers and are created for a specific purpose, like a brewery anniversary or a special beer release. One that was in particular demand was Nashville's Yazoo Brewing Company's black, wide-based snifter depicting the flames of a golden sun surrounding a clear circle, made to celebrate the 2017 full solar eclipse. Just one hundred were made, and they sold out immediately.

American breweries have long tried to stand out in the field with their own glassware. Matthew Cummings, for example, is the brewer and owner of Pretentious Beer Co. in Knoxville and holds a master of fine arts in glassblowing. He creates specialty glassware for his taproom and online purchases. Notably, Sam Adams years ago created what it called "the perfect pint," a glass to be used with its Boston Lager. Vaguely hourglass in shape, it featured a wide bowl atop a concave stem. An outward-turned lip facilitates easier drinking, and laser-etched nucleation points at the bottom of the glass help attract and direct a steady stream of carbonation to the top, advancing the beer's aromas and flavor. It's a nice concept and has quite a bit of science and ergonomic know-how behind it. It was popular for a while at certain spots—the brewery made a big push to get it into large chain accounts like Applebee's, where it was omnipresent after its 2007 release. You'd be hard pressed to find them in bars today (except maybe at a register, holding pens), but you can still see them in the brewery's television commercials and print advertisements. You can also drink from them when visiting the Sam Adams pilot brewery and tasting room in Boston (where they are available for sale in the gift shop, naturally). After a decade of being in the public's eye, it's an immediately recognizable glass, but despite a big marketing push and what Jim Koch said in multiple interviews was hundreds of thousands of dollars spent in research and development, it's still not widely used—which shows the persistent power of the shaker pint.

Some brewers, however, have made stronger inroads with American consumers thanks to glassware. Stella Artois, the lager made by the Belgian-based brewery owned by Anheuser-Busch InBev, pushed into the US market several years ago with a splashy ad campaign, installing branded tap handles at bars across the country, and providing each account with chalices specifically for their beer.

Now, let's be clear. Stella Artois is a fine beer, a humble lager dating back to the 1920s, first brewed as a Christmas offering. It's far from a luxury product—more of a solid workhorse. Because of this, the beer was associated in certain circles in Britain with binge drinking and violence. The brand needed a makeover. By installing those sleek, curved chrome tap handles, training bartenders to do an elaborate little ritual when filling each glass (involving a water rinse and then a swipe across the top of the filled glass with a knife to wipe off foam), and charging a few extra bucks per serving, the brewery had customers thinking they were getting something exceptional. And attitudes changed. The campaign worked, flipping a switch for many people. Perhaps it wasn't the best beer they had ever drunk, but the presentation and pomp made them feel special and turned them into loyal fans. The beer didn't change, just the window dressing.

What this big marketing push did do was drive dollars for the brewery. Stella Artois became an instant success, a top imported beer, toppling other brands that until then had done relatively well, most notably Amstel Light, owned by the Dutch brewer Heineken.

Even so, much like Sam Adams and the perfect pint, you don't see the Stella Artois glasses as frequently as you used to. They show up in commercials and at some bars, but as the Belgian lager has become more commonplace, it's more likely than not served in a shaker pint. Once the initial marketing push wore off, customers still bought the beer, not caring about the glass. Again, this demonstrates the power of the shaker pint. We just accept it.

When Amstel Light—you might remember it as the "beer drinker's light beer"—saw its sales drop, they tried to capture some of the Stella Artois lightning by introducing their own glass to the US market. Similar to the Stella glass, it was a pilsner-type flute on a base instead of a chalice. A revamped marketing campaign moved away from its familiar tagline, and urged people to "live tastefully." Then the brand switched to calling itself "One Dam Fine Bier," *dam* being short for Amsterdam, of course. It never caught on. Customers weren't swayed by the glass or the words, and these days you're hard pressed to find the beer on tap in the country. It was a similar story for Bass Ale (also owned by Anheuser-Busch InBev). That classic British pale ale, with a storied past and familiar logo (its red triangle has the distinction of being awarded the first ever trademark), saw sales decline. A pilsner glass with a triangle-shaped base and the brand name spelled out vertically was introduced. It petered out and has all but disappeared from shelves.

Sometimes marketing and visuals can remake a brand. Other times they are the last things we remember before the brand recedes into memory and is lost to time. Stella was successful thanks to a well-timed and well-funded marketing push. Amstel clearly made missteps in trying to answer the threat, choosing to copy rather than forge a new path or reinforce its standing. These are risks that all businesses run, and it's no different for the beer industry, where sometimes brands simply fade away.

If there's one brewery that benefits more than others from branded glassware in America, it's not Stella; it's Guinness. The famed stout from Dublin has a glass that is part shaker pint, part nonic pint. Recent iterations have the harmonic curve of a harp (long used in its branding) blown into the glass. It perfectly showcases the signature Guinness nitro pour (more on that science coming up), and that's the best advertisement a beer can get. Better than all their classic zoo-animal print advertisements, flashy television coverage, or clever radio spots, seeing a Guinness out in the wild, poured just down the bar from you, makes an immediate impact. Watching the cascading beer rising to a firm, cake-like white head is mesmerizing.

Diageo, the brewery's parent company, knows the importance of good marketing. To build anticipation, they encourage bartenders to take 119.5 seconds (why it's not just two minutes is still unclear) to pour the perfect pint from what is usually a dedicated and well-branded tap. Pull the handle (shaped like a Guinness pint), then pour the stout until it reaches the halfway point. Allow the ale to settle, and then fill the rest to the top, placing it in front of the customer, who can watch the great show that is the cascading rise.

In major markets like New York, Boston, and Chicago, the brewery makes sure bars are well stocked with its signature glassware. Still, it's not uncommon to see the stout poured into a shaker pint, or at certain places into plastic cups. Once, on the eve of a St. Patrick's Day in New York, Fergal Murray, who was then Guinness's master brewer, admitted to me that it was unlikely and unreasonable to expect any bartender to follow strict serving rules. It's all about volume to the customer, he said. His words crystalized a realization for me: no matter how much any brewer wants you to have the full experience, there's nothing about serving beer that's written in stone, and almost nothing bar owners and bartenders are willing to enforce at the expense of losing a buck.

Not that some folks aren't trying. At the Bierstadt Lagerhaus in Denver, they take every aspect of their service seriously. Especially the glassware and the pour. Each of their lagers and pilsners has a dedicated glass, and what you order will be served in the corresponding vessel every time. For their pilsner, they deploy the slow-pour method—which is exactly what it sounds like—creating a little bit of theater and a beer that arrives at your table with a fluffy tuft of foam above the rim. (It's a nice presentation, if impractical for a review tasting.)

The glassware doesn't stay within the walls of the pub either. Co-owner Ashleigh Carter is clearly a student of history and has publicly said that she wants to replace every Stella Artois handle in the greater Denver area with a tap of their own pilsner. She points to the success of Left Hand Brewing Company's nitro stout in knocking off Guinness in the area. One way Carter aims to get customers is from the Stella playbook. Each of the bars—and it's fewer than fifty at this point—that serve her beer must use the correct-branded glassware. No exceptions ever. If the bartenders roll their eyes, or owners complain about the space taken up by the glassware, no matter. Bierstadt will find other establishments that understand their purpose and philosophy. They are staying true to the message they want their beer to communicate.

I wish more places had that conviction. Beer should be celebrated and admired, and consumed in the proper way. Cutting corners for convenience, for a little extra profit, or out of laziness holds beer back from what it is truly meant to be. The shaker pint, like the cockroach, will never die. It'd be great if it did, so in the same way that we reach for the can of Raid when we see a bug, we should also speak up and urge bars and breweries to serve us in better glassware.

That's because the glass we hold can actually help the beer taste better. And isn't that the whole point?

TASTING BEER IS NOT COMPLICATED, FOLKS. PICK UP THE GLASS, SIP, and repeat until it is time to order another or go home. Simple enough, right?

Well, yes. For a long time in our collective beer-drinking experience the beer was just beer, and our imbibing could be done on autopilot. So long as it was clear, yellow, fizzy, and smelled and tasted like we remembered from previous outings, what did we have to worry about?

Now, with thousands of breweries turning out thousands of different beers with an infinite number of flavor combinations, there's a lot we should take into consideration before we tuck into an unfamiliar beer. Luckily, even done "properly," tasting beer is not as intimidating as tasting some other beverages (I'm looking at you, wine). Tastings at vineyards can be churchlike affairs. There's the presentation of the bottle, the show of the uncorking (or untwisting, but that's less alluring), the first pour, the swirl, the sniff, the sip, swish, and then . . . spit? I never understood that. One thing you'll almost never see at a social beer tasting is a spit bucket. Beer was made to be drunk. However grand the theater of a wine tasting, the reverence, deserved or not, afforded to each bottle has led many to think that wine is more sophisticated and complex than beer.

THERE ARE FOUR MAIN THINGS TO TAKE INTO CONSIDERATION when trying a beer for the first time: appearance, aroma, flavor, and mouthfeel. Each plays its part in the overall experience, helping you decide whether you like the beer or want to opt for something else. It can be intimidating to walk into a brewery with a dozen or so taps boasting a wide range of flavors. If you know you like malty, chocolate, or coffee-flavor-forward stouts, then those might be your natural first choice. Same if you dig aggressively hoppy IPAs that lean heavily on tropical fruit aromas. Or maybe you like sour ales that have a strong citrusy, acidic character that others might find bracing. Good. Start there.

The staff at better bars and breweries—that is, ones that take the time to print menus with flavor descriptions or post chalkboards with solid information about the beer—can help you make informed decisions. The peanut butter and coconut porter they advertise should taste just like that, and your taste buds should confirm it. Unfortunately, a lot of places still exist that don't share any pertinent information when it comes to what they are about to pour, so you're left to tease it out yourself.

First, know this: you can always ask for a sample of something before you commit to a full pour. These bars and breweries want your business, and they'll do what they can to make the sale. Just be mindful of other patrons waiting to order if you choose to meander down the whole row of taps before making a selection.

Second: most of the samples will be no more than an ounce, and that's an incredibly small volume on the basis of which to form a full opinion of a beer. You truly need at least 8 ounces of a beer to get its full experience. So, if anything in that small mouthful sparks the pleasure centers in your primitive brain, go for the full pour and try to figure out what the appeal really is. Because remember: if your first impression is that you like it, you probably do.

APPEARANCE

When you judge a beer, the first thing you're asked to consider is usually aroma. That makes sense because our human impulse is to grab a glass, pull it to our face, and take a big whiff. But I argue that appearance is the best first stop. Taking in a beer visually gives us a moment to pause and consider all that is to come, and begin to formulate a solid opinion.

When a beer is poured and served to you, take a look at the color, the head, and even the glass. Assuming the glass is completely clean—no carbonation sticking to the side—let's start with the color. What does it make you think of? Does an amber ale remind you of the color of fall leaves? A hefeweizen of a lazy summer-afternoon sunset? Does a Russian imperial stout, one that sucks away all light, remind you of a cozy winter's evening wrapped in a warm blanket before a fire? Does that citrus-forward New England–style IPA look like pulpy orange juice? Yes? No? Good. Remember: this is personal, but everyone will to some extent associate colors with experiences, places, and flavors. Allowing these connections to occur will put you on the right track.

The color and clarity of your beer can range from a very pale, golden yellow, so clear that you can read a newspaper through it, to a dark, light-grabbing, thick black. The color of a beer is measured professionally as the Standard Reference Method (SRM), and it ranges from two (defined as pale straw) to forty (black). In between lie all shades of yellows, oranges, ambers, and browns.

Beer primarily gets its color from the malt used during the brewing process. Remember back when we talked about kilning? The standard two-row barley malt used in beer can be roasted to different temperatures that bring out different flavors. Some lagers also use a form of food coloring, which helps achieve that very pale color. Adding a grain like rye or wheat can bring more sustentative color or haze to the visual experience. Just because you can't see through a beer doesn't mean it's flawed.

Then there's the beer colored by outside ingredients. You might get a pink beer from a hibiscus addition, or closer to a purple from blueberry. If the beer is pale enough on the SRM scale and features an adjunct ingredient that can add color, you'll see it in the final product. Some beers, like magenta-colored ones, made with fresh berries, can be quite visually arresting and appealing. In early 2018 "glitter beers" started making the rounds. Brewers add swirls of colorful, edible glitter to beers, giving them a magical look—and a social media moment.

Also look at how light plays with the beer. In some cases sunlight might pass through the glass, causing a sun-dappled effect or casting a Stonehenge-like shadow on the bar or table, offering up a good social media moment. In other cases, as with a stout, you might think that the beer captures all the light, but pick up your glass and tilt it ever so slightly. You may be surprised to see that where the beer thins out against the glass, it'll be a lighter color like brown or even garnet. That side angle actually provides a truer representation of the beer's color than the collective hue that is presented in a straight-on view.

Now, if you can see into the beer, what's the carbonation doing? Is it concentrated into tight streams and rising at a rapid pace, like traffic on the freeway? Is it spread out and climbing lazily? We'll dive further into the important role of bubbles in just a bit, but for now take a look and see what they are doing and what their action makes you think of. And know that each little air pocket is bringing the all-too-important aromas to the surface and with each explosion is creating the head (or not) atop your beer.

The head, or foam, also plays an important visual part in the beer experience. The head presents different looks, colors, and textures, depending on the beer. You can have a skin of foam atop a beer, or none at all; other styles will have a fluffy, mousse-like froth above the liquid. Some beers have light and airy bubbles, where others will be thick—like the cake atop a nitro pour of Guinness. As the beer is emptied from the glass, some foam might stick around and lace the inside of the glass (a double IPA is a good style for that). The foam from other beers might disappear quickly. The color of the foam also plays a role in our visual perception. Some heads, like those atop a pristine lager, might be pure white, while more caramel or amber-colored beers will present an eggshell color on top. In most cases, the darker the base beer, the darker the foam. While it's true that a nitro pour from a Guinness will yield a white foam, imperial porters, chocolate stouts, and smoked stouts might have more mocha or tan-colored heads, like you'd see on an espresso.

Next: if you're a student of styles, judge whether the beer looks appropriate for a lager, stout, IPA, and so on. If you're just here to drink some beer, and if it looks good to you, then you're ready for the next step.

AROMA

Smell is the most important part of the beer-drinking experience. What we perceive as taste is actually about 85 percent aroma, meaning that by the time the beer gets to our taste buds, we've already formulated an opinion on what's to come.

Let's assume that the beer has been properly poured, meaning there is ample space between the top of the head and the rim of the glass. This is more important than you might think. It's true that we're so used to glasses being filled to the absolute tippy-top that being served a glass with space to spare might feel like we're being cheated. But having room between the liquid and the top of the glass allows aromas to pool so that when we get our nose close we can actually breathe in the scents instead of taking a bath.

Each ingredient in the beer in front of you contains odor molecules of its own. In a finished beer, they've come together and had the chance to create new compounds. For example, Carafa malt on its own smells like chocolate or espresso. The Mandarina Bavaria hop smells like tangerine. The two together might make you think of a chocolate-covered orange, the kind of candy popular around Christmas. That's the right track to be on when it comes to assessing what your nose is putting together. In some cases the beer might have just one dominant aroma. It's a fun experiment to see if you can tease out which ingredient is strongest.

When you're ready to get started, first gently swirl the beer around in your glass. A good rule of thumb that I was taught is to think of a record player; get the liquid turning at about forty-five rpm. Once the beer is spinning, lean in and sniff. Your instinct will be to plunk your nose into the glass and give one big inhale, taking in as much as you can as quickly as you can. In this scenario I think back to Saturday-morning cartoons, when a big roasted turkey, perfectly browned and radiating steam to represent flavor, is pulled from the oven. The character opens their nostrils wide, breathes in deeply, and then exhales with satisfaction. This method seems like it should work, but in the case of picking up nuances in a beer, it is the least helpful.

After participating in countless beer tastings with longtime professionals and judging beer competitions around the world, I've found that there are two solid and reliable methods to getting the most aromatic information from a beer. The first is similar to how a bloodhound would act: take quick sniffs at the top of the glass, allowing ambient air between each short breath and allowing aroma compounds to build with each iteration. The first sniff might reveal a familiar smell. The second will build upon that, by the third and fourth it's coming into a clearer focus, and soon your brain is identifying the aroma from your memory banks.

The second effective manner of picking up aromas is to put the glass about an inch under your nose, but to the right. Then pass it beneath your nostrils, pulling to the left. Think of a typewriter reaching the end of a page and the platen roller being pushed back to the start position. What happens in this scenario is that you're beginning with ambient air, then a slow, deliberate stream of information is delivered to the brain, helping you to process what's bursting from the glass.

What's happening in your head at this point? All manner of interesting things, actually. The average human nose is lined with up to two million nerve endings that interact with aromas as they enter your nasal cavity. After the scents pass through the nasal cavity, they gather in the olfactory bulb, which is at the front of our brains, set back, if you can imagine, from where the top of the nose meets the eye sockets. Here, the aromas are processed, and they travel to the olfactory cortex for further identification before being sent throughout the rest of the brain, where the perception is given context and articulated to you through a memory. Because the various parts of our faces are so tightly linked, you can actually pick up aromas at the top of your throat by breathing in through your mouth.

The brain has a great storage capacity, so let's assume that right now in your glass is a simple wheat ale that received an infusion of raspberries before it was packaged. Most of us have encountered raspberries before. We know what they look like, what they smell and taste like. We have very specific olfactory memory when it comes to an aroma we've encountered our whole lives, and it can be easy to pick out that scent quickly. "I smell raspberry. This beer must have raspberry in it." Even a bite of the somewhat bitter and earthy raspberry seeds leads to a specific aroma or sensation that is burned into the mind. A whiff of something familiar can bring back old memories, like the time you went raspberry picking as a kid and then stopped at Dairy Queen for ice cream on the way home.

The memory of aromas can also play tricks on us. If you're doing a blind tasting and smelling this beer for the first time, the perception of raspberry might trigger a different response in your brain—say, blueberry. This seems counterintuitive, but because raspberry is so familiar, without seeing the actual fruit we can easily confuse its fragrance with something else, like blueberry, or strawberry, or even cherry or black currant. The same is true should any of those other fruits be the dominant ingredient: we might pick out raspberry in its place. And once you've formed an opinion on what the flavor is, it can be hard to expand beyond it.

There's also the heavy power of suggestion. If we are told that we should expect a certain flavor, our minds will naturally gravitate toward that idea. If I hand you a glass of raspberry wheat, but say, "Here, try this strawberry ale," your brain will go straight to strawberry and you might not even consider searching for another flavor. If you do go hunting (and we always should try to go deeper into the aroma of each beer), your brain might tease out raspberry, but because I've already said strawberry, you doubt yourself. This is all too common, and something that professional tasters must try to get past.

When I was editor of *All About Beer Magazine,* I regularly held blind review panels at my home. Each of the judges would be given a sample pour of a base style, say a porter, and if it had a special ingredient added, like coffee, I would only tell them that it was a "porter with something else added." It was amazing to listen to the panel reveal their tasting notes. Rather than identifying coffee, tasters would talk of bell pepper. The two foods share the compound 2-isobutyl-3-methoxypyrazine, and for many people the aroma of coffee makes the brain default to identifying it as pepper. I always found it curious that even repeat panelists would first assume bell pepper rather than coffee, even though pepper isn't a common ingredient in beer, but java is. Now when I'm tasting and I come across green pepper in the aroma, I force my brain to think about coffee. It's like a light switch flipping on: there it is.

We go a step further on *Steal This Beer,* the podcast that I cohost with Augie Carton of Carton Brewing. A guest brings a beer for us to taste, and we aren't told anything about the style, the ingredients, or even the maker. It's served to us in black glasses, so we can't see the color or the carbonation. All we can do is smell and taste and consider mouthfeel. Sometimes we're able to hit the nail on the head, pulling out individual ingredients, identifying hop varieties, and even naming a specific beer—especially if we've had the beer multiple times in real life. Other times we're wildly off, like the episode when we both swore that Machine Czech Pilsner by Bunker Brewing Company—which is very light in color, contains Pilsner malt, and is lightly hopped—was a rich milk stout, dark in color. Also, one is a lager and the other is an ale, and one has lactose sugar in it and the other does not. But a bit of chalkiness in the lager, plus a thick mouthfeel and a touch of sweetness, put us on a path to thinking it was something else entirely. It was a humbling but teachable moment. The previously mentioned Orval regularly makes appearances on the show. Each time it vexes me because of how it ages. Sometimes the bubblegum yeast flavors will dominate; at other times the herbal hops do. But nearly every time someone brings it on the show (and guests do so just to mess with me), I find something different and have a hard time identifying it.

Playing with aromas is a fun sensory experience. At certain brewing schools they will blindfold students, hand them an apple slice to bite, but hold an orange under their nose. The overwhelming scent of the citrus will convince your brain that you're eating an orange slice rather than the fruit of Eden. This is a good experiment to perform at home because it shows how easily we can be led by our sense of smell. When cooking (or homebrewing), take a moment to smell each ingredient, even familiar ones like salt and pepper. Allow your mind to wander as you inhale, see what the scent sparks, and know that somewhere in the deep recesses of your brain this new olfactory memory is being saved.

FLAVOR

After all this looking, swirling, sniffing, it's time to drink the beer. Go ahead and take a sip. When trying a beer for the first time it's helpful to follow the Goldilocks rule: not so much that your cheeks puff out and you need to gulp it down to catch your breath, and not so little that you could open your mouth without any liquid falling out. Somewhere in the middle is just right. Enough liquid to submerge your tongue if it were pressed against the bottom of your mouth will do. This amount will help your taste buds get down to work identifying the flavors and the various qualities, like bitterness, sweetness, saltiness, sourness, and umami.

Our taste buds are essentially bumps on the surface of the tongue that perceive qualities in the foods and drinks we consume. The bumps are known as papillae, and we can see them if we stick out our tongue while looking in a mirror. The taste buds lie inside the bumps you see, and they interpret various flavors and then send messages to the brain with the pertinent information. Our taste buds live for about two weeks before being replaced, although actions like scalding your tongue with hot coffee also wipe them away. As we get older, especially entering our forties and fifties, we grow fewer taste buds, and by the time many of us hit our sixties it can be difficult to accurately tell the difference between various flavor qualities.

Quick aside: you might remember something called a tongue map from your old schoolbooks. This diagram divided the tongue into zones where various flavors were supposedly detected. That idea has been roundly disproved over the last few years; scientists now agree that the entire tongue can be an accurate receptor for all the different flavors.

Let's discuss the flavors, shall we?

Bitterness is generally defined as a sharpness or identifiable lack of sweetness, and it's something that most of us are sensitive to. It's one of the reasons why hops, which can be bitter, are such a polarizing ingredient for beer drinkers.

Sweetness is a sugary sensation that most of us associate with happiness, thanks to an early introduction to candy and other treats. In beer, malt and yeast can both impart sweetness, along with the introduction of any number of specialty ingredients, including, you guessed it, sugar.

Sourness is represented by acidity. When it is perceived on the taste buds it can cause the salivary glands to kick into overdrive. A number of beer styles celebrate sourness, like gueuze, which uses wild microbes and takes time to develop. Then there's the recent rise of quick-kettle sours, such as Berliner weisse, which use *Lactobacillus,* a genus of bacteria that work quickly to convert sugars into alcohol and impart a lemony, citric-acid quality. Sourness can also come from the addition of citrus, like lime juice, to a recipe. (See my discussion of sour beers in Chapter 3.)

Saltiness comes from, yup, salt. In beer the quality of saltiness can be introduced in a number of ways, from the brewing water, which might have some salinity to it, to the addition of pure salt. Gose is a German style of beer with a savory character, thanks to the addition of salt during brewing. Contrary to what you might experience if you were to drink ocean water (not recommended), this beer style is usually lighter in alcohol and quite refreshing, in much the same way that some folks find brine-based drinks to hit the spot. Salt is a natural accompaniment to beer, as it can act as an enhancing counterpoint to other flavors, especially sweet ones. A few years back a spate of salted-caramel beers hit the market (mimicking a popular ice cream trend at the time), and the combination is a clear winner to anyone who's had it. There are even parts of the country, namely the Dakotas, where it's common for drinkers of a certain age to sprinkle salt from a shaker into their glasses between sips to give otherwise sweet American lagers a bit of a boost.

Umami is a savory quality that speaks of age and depth. Most people encounter it in fermented foods like soy sauce or kimchi, old cave-aged cheeses, or a suitably aged steak. In beer it becomes apparent in higher-ABV beers that have been properly cellared, such as barleywines, or in barrel-aged wild ales that begin to take on a soy-like character.

MOUTHFEEL

THE LAST THINGS TO CONSIDER WHEN TASTING A BEER ARE BODY and carbonation: the two main components of a beer's mouthfeel. Assessing mouthfeel is as simple as taking a sip—of the size we talked about above—and holding it on your tongue for about three seconds. While doing this, think about two things: how the beer feels, weight- and texture-wise, and the level of carbonation. Keep in mind that although it might be natural to assume that dark beers are weightier, color is not always directly correlated with body. Each beer is different, each style has its own specs, and knowing as much about a recipe as possible can help us decide if the beer is truly well made or masquerading.

First let's tackle body. I hope it's a fair assumption that you've had a Guinness at some point in your life. It's one of the world's most popular and recognizable beers. Thanks to its nitro pour, when it arrives at your seat it looks thick and inviting. Take a sip and hold it. Feel the liquid in your mouth, and move your tongue around a bit. You might be surprised to realize that it has the consistency of a glass of water (plus carbonation, of course). Despite its dark and thick appearance, Guinness is actually quite light and mild in body. That's one of the reasons it's such easy drinking. Although it's in the same family as a Russian imperial stout, it is nowhere near as heavy, viscous, and assertive as those beers tend to be. That's because of the ingredients used to create it. I consulted *Craft Beer and Brewing Magazine*'s homebrewing recipe archives to get a handle on just what makes these two beers feel so different, and, of course, it's the malt.

Malt not only gives color to beer, it also contains the sugars that yeast feeds on to create alcohol. The more malt in a beer and the higher its sugar content, the boozier and thicker the beer is likely to be. For a five-gallon homebrew batch of dry Irish stout, a recipe called for 6 pounds of pale malt, 2 pounds of flaked barley, and 1 pound of roasted barley. That equals a dark ale with some mild chocolate and coffee flavor, but mostly the character of light cereal grain and a slightly nutty taste. Five gallons of an imperial stout calls for 19 pounds of two-row malt, 1½ pounds of roasted barely, 12 ounces of chocolate malt, 8 ounces of black patent malt, and 16 ounces of two separate kinds of caramel malt. The larger quantities and greater variety of malt lead to the thicker mouthfeel. If thick beers aren't your thing, but you like the taste of coffee and chocolate in your beer, a dry Irish stout shouldn't be intimidating.

As I've already said, it's not just darker beers that are thicker. Hops can lead to a robust body thanks to the oils contained within them. A double IPA, for example, will have a thick and oily character to it, and the hops will make the beer feel chewy, a bit like weak paste in your mouth. Wheat-forward yet light-colored beers, like hefeweizens, also have more body than a golden-colored blonde ale that uses lighter malts.

Carbonation is important in assessing a beer's mouthfeel. Again, going back to the example of American light lager, we've been conditioned to think that our beer should be fuzzy and effervescent, offering a light prickle of carbonation on the tongue but not harsh enough to be unpleasant. And that's generally true; for most styles you want the lively kick of carbonation. But not every style calls for the same uniformity.

Cask ale, with its natural carbonation, will be softer on the palate. The same is true for nitrogenated beers. Other styles like barleywine or the above-mentioned imperial stout should have some vibrancy, but not the prickle of a lager. Saisons and other bottle-conditioned beers, especially if they are in a cage-and-cork bottle, will likely pop upon opening with the robustness of Champagne and will send up the bubbles to match. There are other styles, such as the salty gose, that are so carbed up it's almost like having those little scrubbing bubbles from the television commercials dancing on your tongue.

We focus a lot on taste when picking a beer—do I feel like something bitter, or something dry?—but the more familiar you become with all that beer has to offer, the more you'll realize that mouthfeel is almost more important for matching a beer to your mood. Sometimes the occasion calls for effervescence, while other times you'll crave mellow. As you pair your meals with beer, you'll find that the different textures in your glass can wonderfully complement or contrast with what's on your plate. The ultimate goal is to give each beer its fair shake and let it reveal its true nature to you.

Now consider the beer in your glass and put it all together: the look, the smell, the taste, and the mouthfeel. Did you like it? Did the beer meet the promises of the style and give you a good sensory experience along the way? Has it made you smile, or just become a background accompaniment to your environment and activity? What do you like about it? Would you order another one? If you're answering yes to any or all of these questions, congratulations! You've found a beer you like, one you can keep coming back to for several rounds if you so choose.

If not—if you just didn't like the flavor, or found the appearance off-putting, if the carbonation is too rough on your tongue, or the aroma unpleasant—don't worry. Beer is personal to each drinker. Think of what you prefer when it comes to wine. Some folks gravitate to red over white, or vice versa. You might like dry, or sweet, or styles that come from a specific region.

Consider beer in those terms. If you like coffee in regular life, maybe seek out a beer with those flavors. The same goes for fruits, spices, and any of the other naturally occurring tastes in the world. If you think you don't like beer, or if you have a friend who claims they don't, start with flavors that you *do* like—you can find almost anything in a pint of beer. We've established that there are thousands of breweries making thousands of beers here in America. Surely one of them will appeal to you. It's just a matter of seeking it out and giving it a try, and that's a great deal of fun.

WHERE BREWERS HAVE THE MOST FUN IS IN THE ADDITION OF flavors, and this can be both a good and a bad thing. If there's an item or ingredient on this planet that humans can eat, there's a strong likelihood that brewers have made a beer with it. For generations brewers relied on the core ingredients—water, malt, hops, and yeast—to impart flavors to their beers. Depending on the variety of specific ingredients (malt or hops, say), or how those four ingredients were prepared, the resulting beers could taste like coffee or tropical fruit, bubblegum or spice. Now, however, some brewers add those actual flavoring ingredients, and many more, to their beers. They do so for taste, or to stand out, and because brewers are a creative bunch, some do it just to try new things.

Although it might seem easy to brew a chocolate-milk stout by simply adding chocolate to a stout made with lactose (the type of sugar found in milk), in reality it takes strong technical knowledge to incorporate the ingredients properly and ensure that the finished beer is a success. It also takes a good amount of imagination and skill to develop a beer that uses unusual ingredients. If an open-minded brewer has a good palate, creative freedom, experience working with food, and is compelled to make something new, high-quality beer will happen. It's why we now have beers that taste like lemon meringue, strawberry shortcake, or a margarita and that incorporate actual citrus, or graham crackers, or celery salt and horseradish.

The inspiration can come from anywhere. Years ago I talked with Jim Koch of Samuel Adams about their celebrated chocolate bock, available each December as part of a holiday variety pack. There were challenges in getting that beer to the market. After the initial idea, the brewers needed to figure out what kind of chocolate to use—would it be cocoa beans, a finished chocolate bar, or a form of liquefied chocolate? Next, how would it be added to the beer—in the mash, to the boil, or during fermentation? Could other flavors enhance the fresh chocolate, bringing out additional aromas and taste sensations—yes, vanilla, but do you use beans, extract, or puree? Possible combinations seemed endless. Koch says that his team went through a lot of trial and error before finding a recipe that worked. The resulting formula uses cocoa nibs in a long, slow cold infusion to bring out the desired flavors.

Brewers must go through a similar testing period with every new ingredient, from spices to fresh fruits to vegetables. Not everything a beer-maker comes up with will hit the mark, or even pass the sniff test. Brewers have told me of bacon beers, chicken ales, and banana concoctions that still haunt them.

"You can brew with anything," Koch told me. "But does it make a pleasant drink?"

From Buddha's hand (a variety of citrus that has been described as the "Edward Scissorhands of fruit") to wasabi, from black tea to smoked peat, finding beers that incorporate local flavors, like hot chicken spices in Nashville, or fenugreek in India-made beers, gives us a sense of how people live around the country and around the world. And because brewers are turning out new beers with ingredients sourced from every continent at breakneck speed, some are even looking beyond terra firma and into the skies.

There was a reoccurring sketch on the original *Muppet Show* called "Pigs in Space." I remember watching this *Star Trek* parody as a kid and laughing at the absurdness of it all. Over the last few years I've had a similar reaction as brewers have looked into the great heavenly expanse and wondered what it could do for their beer. No strangers to adding odd ingredients to their recipes, Dogfish Head Craft-Brewed Ales kicked off the space-ingredient arms race in 2013 with Celest-jewel-ale, a small-batch Oktoberfest brewed with a tiny amount of finely ground lunar meteorites. Brewers steeped the lunar dust—in a tea-bag-sized container—into the beer during fermentation, then served it as a pub-only offering in koozies made out of the same material used in astronauts' space suits. The brewery was gifted the lunar dust by ILC Dover, the company that made space suits for the Apollo program.

The beer was more Oktoberfest than moon juice, with traditional flavors of toasted, doughy caramel malt, some herbal hops, and a crisp finish. Personally, I think the brewery should have used some aged hops, which give off isovaleric acid and create a cheesy aroma and flavor. The moon dust itself imparted practically no flavor; however, using such a small amount was likely for the best, since any quantity of significance would likely cause intestinal irritation.

Now, before you head out to your local homebrew shop and ask for a few micrograms of lunar dust (assuming you can afford it, big spender), know that it's incredibly hard to come by. Fewer than two hundred lunar meteorites have been found on our planet. The ones brought back from the Apollo program are "national treasures" and are not allowed to be sold or given away.

Brewers began wondering if there was an easier way to get "space flavor" into their beer. Just a few months after space-gloved hands clinked glasses in Dogfish Head's pub, on the other side of the country, Ninkasi Brewing Company of Oregon traveled to the Nevada desert, loaded a vial of yeast into the cone of a rocket, and blasted it into space. The hope was that the yeast would remain viable, would possibly take on some new flavor properties (or superpowers from radioactivity in space?), and retain the ability to create beer back here on planet earth.

It was a spectacular launch on a clear morning just after daybreak. The rocket's trip took just twelve minutes. Since yeast is delicate and susceptible to damage in harsh conditions, it needed to be recovered and repacked on ice within ten hours of its return to the ground. Unfortunately the radio signal from the rocket was lost on reentry, and searching a potential area of a hundred thousand square miles was as difficult as you'd imagine. By the time the payload was discovered twenty-seven days later, the yeast had died.

Taking a page from our own space program, Ninkasi was not deterred. Several months later they returned to the desert and shot a new rocket into space—77.3 miles up, to be exact. This time the yeast returned safely and was used to make Ground Control, an imperial bourbon barrel-aged stout. For the record, it tastes exactly as described.

At the bottom of the planet, an Australian brewer is looking to the day when space travel will be common for folks like us. They are working with a local aeronautics company to develop low-carbonation, high-flavor beers that can withstand the conditions of space. The hardest part will be developing the right packaging, so don't expect any glass bottles or cans. Think back to childhood and those foil packs of astronaut ice cream. It'll probably be like that.

Unless you have special access to NASA, or are hiding an alien in your backyard shed, brewing a space beer at home will be more challenging than some of the others highlighted in the book. Remember, the yeast that Ninkasi used was only shot into space once and then propagated again and again back here on terra firma. The Dogfish Head beer was made in such limited quantities that only a handful of people were able to try it.

So, it might be best to grab a few of your favorite beers, head into the night, and tip them back while staring at the stars. Consider the universe, your place in it, and think of some more earthly ingredients to add to your next homebrew.

Enjoyment of beer is not just limited to the contents of the glass—it's about environment and context too. For the longest time if you wanted to drink beer you had two obvious choices: the bar or home. Now, you can drink directly from the source, and I actually find that I'd rather spend time drinking at a brewery: testing their lineup, finding one I like, and then whiling away an afternoon or evening chatting with friends or family. Or going in alone, of course. Breweries are generally friendly places, without the same trappings of a bar or nightclub, where anyone can sit by themselves and not be bothered if that's their wish. I've made lifelong friends at breweries, and chances are you will too. Brewery taprooms have become the new taverns and social spots in a lot of neighborhoods. Consider the Nielsen TDLinx study from 2015 that revealed that more than twelve thousand neighborhood bars closed between 2004 and 2014 in the United States. During the same period, several thousand breweries opened across the country. Those numbers are a point of contention for some state tavern and restaurant associations. They regularly chafe at what they see as loopholes that allow breweries to serve beer by the pint without having to buy a full liquor license, which is more expensive in most jurisdictions. That's money that could go to their members, the associations say. Purchases from the food trucks often found in brewery parking lots, especially those without kitchens, hurt their members too.

In the early days of the current beer movement in the United States, most of the breweries that opened had a restaurant attached to them. These brewpubs were, by and large, restaurants first and breweries second; they highlighted their quality food to attract customers, who they hoped would also try the beer. Then the switch flipped, and as a new generation of brewers became surer of their talents—and even surer that they didn't want the headaches that came with operating two separate businesses under one roof—they opted to focus on the beer and outsource the rest.

The results were taprooms that were extensions of the brewers' and owners' personalities. The early brewpubs mostly had an inoffensive British-pub feel, or mimicked a seafood house, or had an old-fashioned American diner vibe. The menus offered something for everyone because they wanted everyone to come in. Now there are breweries that bill themselves as places for heavy-metal music enthusiasts. There are ones for photography nuts, for folks who are way into yoga, others that celebrate the acid-heavy days of the Summer of Love, and those that cater to just about every conceivable ethnic group, from Thai to Indian to Polish to Korean.

This is a strength: wine is not like this. When was the last time someone went to a wine bar to connect with their community? The relaxed nature of beer, the conviviality, the open-mindedness are selling points. It's the opposite of snobbery, in fact. The brewing industry can take what wine is good at—celebrating itself, taking its liquid seriously, paying attention to glassware, temperature, flavors—and leave the rest behind.

Here's the thing: anyone is welcome at a brewery. The way the modern brewing industry has evolved has resulted in very little pretention overall. Visiting a brewery is still just a good, fun time. Sure, you're going to encounter some folks who take it more seriously than others, or who might try to hold their knowledge over other people, but those experiences are still pretty rare, especially compared to visiting an ultracool cocktail lounge or high-end winery. I've struck up great conversations on my visits to TRVE Brewing in Denver, one of the above-mentioned heavy-metal breweries. My bow tie and blazer stood out a bit, but I never felt uncomfortable. I just couldn't hum along with any of the songs on the stereo. For the regulars, however, the ones who gather for pints and conversation under a shared passion, the brewery is a welcome spot and part of their community fabric.

If you're lucky enough to live in a city or area with several breweries, you're likely going to find one that suits your taste, be it a sparse, elegantly decorated space with an always great playlist, or a more muted and subdued location where you and your book (this book, hopefully at least once!) won't be bothered. So long as you enjoy the beer in your glass, you can bet the owners enjoy having you there.

It used to be there were three things you would never discuss in a bar: sex, politics, and religion. Today's climate makes that old (and probably wise) adage almost impossible to stick to, and beer especially encourages conversation because it brings together people of all stripes, persuasions, and opinions. Breweries have become places where freedom of expression is encouraged and where people try to solve the world's problems, or at least try to see other points of view. I've watched an atheist have a civil discussion with a pastor over glasses of saison. I've seen Republicans and Democrats clink pints toward the promise of a better country—for everyone. From talking with people at breweries, I've picked up interesting tidbits, learned more about a foreign city, and even gotten recommendations for the next TV show I should binge watch. Beer brings out community and is also a social lubricant. Maybe you've experienced this yourself. There's something about the brewery setting that is conducive to conversation, much more so than a tavern, the local TGI Fridays, a winery, or a hotel bar.

Even religious groups have started to hold meetings at breweries. As some denominations look to add to their ranks and keep pews filled, events that combine beer with Bible studies, sometimes called "Religion on Tap," have popped up across the country, inviting people to sip and talk about lessons from the good book and ways to improve their lives.

One phenomenon that's generally being lost to brewery success is the "mug club." In the early days of the beer revolution, breweries—usually brewpubs—charged regular customers annual dues in exchange for having their own dedicated mug. Usually these were ornate or elaborate vessels engraved with the member's name or membership number. Members would get discounts on beer, invitations to special events, access to special bottle releases, and often a larger pour than a pint. As breweries blossomed, so did the number of people who wanted to join the clubs. Soon enough, most became unwieldy, and the majority were phased out. If the trend is ever widely revived, I'm still on the waiting list for the one at my hometown brewery.

Mug clubs have been replaced by bottle societies. These membership-only ventures ask patrons to fork over cash in advance of special bottle releases. The promise is that the member will get access to exclusive beers unavailable to the general public, or at least a head start on ordering them. While obviously designed to taste great, these beers come with a high price tag and mainly offer bragging rights. The breweries that offer bottle clubs usually have a solid reputation, so customers expect their special-release beers to be of exceptional quality, and members don't mind shelling out cash for a special surprise. Wait lists abound for the more celebrated of these clubs.

In addition to creating communities *among* their customers, breweries tend to be actively involved in supporting the communities that exist *outside* their walls. If you haven't recently, the next time you visit a brewery take a moment to admire the walls. The larger the brewery, the more space they have to hang things. Many breweries have become art galleries—both in the traditional sense, in which a rotating crop of local artists can display their work in the hopes that someone will buy it, and in the more modern sense of supporting street art, with murals painted on brewery walls. A lot of fantastic local art is being produced from spray cans. Some depicts life in a certain region or culture, or is simply rooted in pop culture. If you're in Asheville, North Carolina, check out the bizarre yet always Instagram-worthy mural portraying Tom Selleck hugging Sloth from *The Goonies* on the back wall of Burial Beer Company. In Miami, Star Wars–inspired paintings form part of the decor at J. Wakefield Brewing. Then there's the mural created from Post-It notes of the late Michael Jackson (the singer) holding his pet monkey, Bubbles, that clings to the wall of Modern Times Beer in San Diego. You can learn a lot about an area (and a brewery) by checking out the art the breweries choose to display, even if it's not for sale.

When Jack McAuliffe opened New Albion Brewing in the 1970s, the town of Sonoma, California, didn't know how to properly license him. The city showed almost no interest in helping him with his small business. Today politicians bend over backward to get the brewery vote at both the local and state levels. Even former President Obama tried his hand at diplomacy over a few "beer summits," and he brought homebrewing back to the White House. You're hard pressed to find a ribbon cutting these days at a brewery that doesn't have a mayor posing for pictures or glad-handing the crowd.

It's not uncommon for brewers to hold political forums, inviting politicians to discuss platforms and issues, or weighing in themselves. When it comes to environmental and social concerns, brewers are mighty vocal and can use their considerable clout with customers to get results. Some brewers have successfully run for office. Governor John Hickenlooper of Colorado started Wynkoop Brewing in downtown Denver in 1988.

Breweries often encourage or even sponsor charitable-giving campaigns. Following natural disasters, you'll find many breweries urging customers to donate canned goods, clothing, blankets, or other supplies to be distributed to people in need. Some places even offer free pints as thanks. You can drink beer *and* do good.

We take all these factors into account when thinking about where we should spend our limited time. Breweries are interested in making sure their patrons feel comfortable spending both money and time at their establishments, whether it's twenty minutes for a quick pint, or five hours scrolling through their phone or catching up with friends. Each time you return to a familiar spot, you're supporting a small business, and ideally getting something in return. If the beer is to your liking, the glass clean, and the ambiance comfortable, there's every reason in the world to return. And if not? There's likely another spot just down the road.

Regardless of what kind of beer drinker you are—casual or super committed—there is so much more that comes with having a pint these days than just having a pint. When I started drinking beer back on my twenty-first birthday, I never imagined I'd be exposed to such a vast world. Going beyond the glass, I've been introduced to new ingredients and flavors, new people, new concepts, music, philosophies, and places—all thanks to beer. By welcoming the adventure, I've found it's possible to live a fuller life.

FIVE

DRINKING AT THE BAR

LET'S GO BACK TO OUR MIND PUB AND ORDER ANOTHER ROUND. Got it? Hopefully what you ordered is in the glass of your choice—even if it's a shaker pint—and that glass is clean. I'll take a guess here that most of you are drinking beer that came from a tap.

The draft, or draught, system is the most common beer-dispensing method in the United States (although perfecting it remains a challenge for many establishments; more on this below). In the grand narrative of beer, draft beer is fairly new. It was not until the late 1700s that the beer engine was introduced to pull beer from a cask into a glass. Before that, the fermented beverage was poured directly into a mug, quickly losing what natural carbonation existed. Advances were made to mechanical CO_2 systems about fifty years ago, finally allowing drinkers to get carbonated beer on draft that could mimic beer from a bottle, which offered a carbonated beer as far back as the 1800s.

There are three general types of draft systems—temporary, direct-draw, and long draw—each with its own pros and cons. Each has a number of components, including hoses, connection points, and regulators, which play a crucial role in making sure the beer is presented the way the brewer intends.

The temporary system is the preferred choice for the casual backyard BBQ or even some beer festivals where drinkers use a picnic pump. The portable pump forces compressed air into a keg, which in turn forces the beer to dispense from a thumb-operated tap handle. Some models can accommodate single-use CO_2 canisters. More sophisticated brewers or breweries use a jockey box: the CO_2 pressurizes the system, and the beer is brought to the desired temperature by flowing through an ice chest before being served. It's usually contained within a specially fitted plastic beverage cooler and is easily transported from one spot to the next with minimal setup.

Bars have a more permanent solution in place. Once a keg is delivered from the brewery and put into place, it's connected to two tubes. The first leads to the tap, and the second to a CO_2 tank that helps push the already carbonated beer out of the keg and through the lines, while also filling the empty space in the keg to keep any remaining liquid in good condition. When the tap handle is opened, the beer is dispensed. This is the setup used for both short-pour (or direct-pour) and long-pour systems, and is a much better option than temporary pumps. Short-pour systems can include kegerators or other refrigeration equipment (e.g., a walk-in cooler or converted refrigerator) located relatively close to the tap handles. The shorter the draft lines, the more likely the beer is to arrive in good condition. I've seen some lines that are an impressively short three feet from keg to tap.

The less desirable long-draw draft system requires the beer to travel a greater distance from keg to glass. This not only increases the potential for contamination; it also causes cooling issues and beer loss. I've encountered lines that run for several hundred feet and cover multiple stories of a building (usually from the basement up). The longer the line, the more chance for problems. A short pour means less time in the tubing. If you have beer running for several hundred feet, there's a strong probability that the beer in your glass has been sitting inside the tubing between pours. Hopefully the line is insulated, or glycol jacketed, to protect it from the effects of temperature changes, but that's not as common as you'd expect. The beer can also lose carbonation, its vibrancy, while hanging out in the tube. If you know of a bar with long draft lines, take care not to get the first pint of the day, because unless the lines were flushed, you'll get a beer that's been sitting out overnight.

Long draft lines also are less likely to be cleaned thoroughly and often, and even less likely to be replaced often. Quite a few beer bars, realizing the importance of clean lines, will change them out between kegs, or at least flush them with sanitizer. This is especially important when you're replacing a line of, say, smoked porter with a lager. Or a wheaty hefeweizen with a gluten-free beer. Some states mandate that lines be cleaned at least once a month; other states require an even shorter time frame. Cleaning and replacing lines is usually done by beverage distributors or companies that specialize in cleaning lines, although most bar owners and bartenders know how to perform the task and regularly do it themselves. The task is labor intensive and usually conducted outside the public eye, because it's impractical to have a draft system exposed during operating hours. For a customer, the proof of a clean line is not even thinking about the system after you've taken your first swallow. I'm always amused (but appreciative) when I visit a new bar and they brag about how short their lines are.

A relatively new draft system called Rack AeriAle pulls beer directly from wooden barrels and carbonates it with nitrogen and CO_2 on its way to a standard nitro tap. This type of system is overdue but is gaining market share, thanks to the number of barrel-aged beers that are being made these days. Expect to find it only at breweries for now, because it would be awfully expensive and risky for a brewery to let a flavorful and reusable barrel out of its protective custody. It's a fun experience to have a glass of beer pulled directly from the wood.

There are times when bars and breweries let their hair down, so to speak, and experiment with draft beer. You may have, on occasion, come across a Randall sitting on a bar or close to the taps. It's a tube-shaped contraption stuffed full of ingredients that can range from chocolate to fresh fruit to raw oysters or fried chicken. Originally designed as a vehicle to deliver additional hop flavor to India pale ales, it was developed by Dogfish Head in 2003, and although it's still used for hops (especially during the hop harvest each late summer), it has evolved into a device to create a fusion of food and beer. Beer geeks—especially those that seek out new experiences on a never-ending quest to expand their drinking horizons—have long celebrated the technology, but more and more the Randall is going mainstream, showing up at traditional spots. It allows bars some creative license with an existing product and gives patrons something new to try, if only for a pint.

IF YOU ARE DRINKING A DRAFT BEER, IT ALMOST CERTAINLY CAME from a keg. Kegs in America come in a variety of sizes, ranging from 5 gallons to 15½ gallons. The two most common sizes that we encounter (at either a bar or a backyard party) are the half barrel and the sixtel, sometimes called a log. A half-barrel keg holds 15½ gallons, or 124 pints of beer. A sixtel holds 5.16 gallons of beer, or about 41 pints. Homebrewers usually favor the cornelius keg, which holds 5 gallons, or 40 pints, and was originally developed for soft drinks. It has a ball-lock valve on top, which can be easier to hook up to a tap or draft system than the aforementioned styles, which need a special keg coupler.

Stainless steel is the preferred material for professional kegs. It's sturdy, durable, keeps beer cold, and doesn't affect the flavor of the beer. Sometimes a keg is wrapped at each end in durable rubber (or completely coated in the substance) to prevent damage if it's dropped or during transport. Kegs get bounced around quite a bit; rubber-coated ones are designed for a lifetime of use. First, a brewery cleans and fills the keg. The filled keg is transported to a distributor, where it's stored, then delivered to a bar, where it's served. After it's emptied, it's picked up by the distributor and returned to the brewery, where the process begins again. Keg washing, by the way, is a hard, smelly, and thankless job. It's also the position that almost every professional brewer begins in.

Plastic kegs were popular several years ago and are still in use. They are made to look like traditional stainless, but they are a cheaper alternative, are inherently dangerous, and really should be avoided. Kegs get beat up quite a bit in the brewery and during transport, which can cause stress on the material. As a keg is being cleaned, it's filled with compressed air, and carbonation from the beer itself can cause pressure. On rare occasions, a stainless steel keg can buckle from the pressure and split along the seams. An overfilled plastic keg, however, can explode, sending shrapnel flying. A brewery worker in New Hampshire was killed several years ago by a piece of debris from a ruptured plastic keg. Some brewers keep pieces of exploded plastic that have lodged into walls and ceilings as a reminder that doing things on the cheap can have consequences.

That said, single-use plastic kegs are available that pose less risk. Think of the "party pigs" that occasionally pop up at a summer cookout. Today, some breweries use similar systems when sending small quantities of beer out of state or to far-off markets, where it's unlikely their stainless kegs would be returned quickly, if at all. These kegs are also recyclable.

BACK TO THE MIND PUB FOR ANOTHER ROUND. WITH EVEN A casual glance we can see color and clarity and hopefully a nice head of foam on the glass of beer. Now, go in for a closer look, and take in the action that is happening inside your glass.

It can be easy to forget about carbonation in beer, unless it asserts itself by, say, fizzing up your nose or aggressively dancing on your tongue—or being conspicuously lacking. Sure, when light filters through a glass, the tiny bubbles can look beautiful as they flitter from bottom to surface. But to think of carbonation as merely bubbles in a glass is to dismiss the science, passion, and countless hours brewers and researchers have put into making sure the effervescence is all it should be. Carbonation is the spark of beer. It delivers aroma. It is effectively stirring the beer while you drink. It contributes to mouthfeel, and its presence (or lack thereof) can help to establish a beer in its proper category.

There is evidence that the ancient Sumerians had foam in their beer, indicating the existence of natural carbonation, which first occurs during the fermentation process when yeast absorbs the sugar in the wort, creating both alcohol and carbon dioxide. But it would be a few more centuries before airtight commercial bottles allowed drinkers to enjoy stronger carbonation one pop at a time, and then a few years more before advances in molecular science allowed brewers to force carbonation into beer.

Carbonation levels are measured as volumes of CO_2. This

measurement is a relative description of how much gas is dissolved in the beer. Volumes vary by beer style, but the majority fall in the 2.1 to 2.8 range—meaning that for every 1 unit of liquid, there are 2.1 to 2.8 units of CO_2 dissolved in the liquid. Cask ales offer less, and some German varieties, like weissbier, and Belgian styles, like lambic, offer more. The ubiquitous American light lagers, as a point of reference, carry a volume of about 2.5. In addition to the naturally occurring carbonation, modern brewers dissolve additional carbon dioxide (CO_2) in the liquid. Some brewers add extra sugars or additional yeast to unpasteurized bottles of beer, allowing for a secondary fermentation, thus introducing additional CO_2. The volumes are measured in pounds per square inch, or PSI, a unit of pressure. Pressure is a combination of quantity and the effects of temperature. When the pressure is relaxed by opening the tap (or bottle or can), the CO_2 separates from the liquid as small bubbles, causing the beer to fizz, form foam, and release the pleasant malt and hop aromas from the key ingredients.

For brewers—both professionals and homebrewers—understanding the science behind carbonation and other aspects of brewing is crucial and often a point of passion. Knowing a little of the science is also important for us, the drinkers, even if it remains in the background of the beer experience. As the beer industry has pushed the ball forward in terms of flavor, mindful brewers make sure that beers with special ingredients are still hitting the desired carbonation levels for the desired styles. Appropriate pressures are something we now take for granted, but consider that not too long ago manuals instructed bars, and even homebrewers, to set the CO_2 for all kegged beers at 20 pounds per square inch and then shake the keg to agitate the beer a bit. Simply put, that was a weak and downright wrong solution that didn't take into account the nuances of styles, leaving some beers undercarbonated and some too aggressively carbonated. Thankfully this is not an issue today. These days, the existence of well-respected manuals, easy-to-use equipment, and a greater respect for the process overall help brewers ensure that the carbonation in their product is how it should be. These measures keep us, the customers, from getting a flat or overly carbonated pour, and keep bars from wasting the liquid that brewers worked so hard to make.

But not all carbonation is created equal. There are three different ways that bubbles get into your beer: natural carbonation, forced CO_2, and nitro. During the brewing process, not long after yeast is added to chilled wort inside a fermentation tank, the fungi get down to work. They begin to ingest the fermentable sugars, converting the beverage to alcohol and releasing CO_2 in the process. This is the natural carbonation. Fellow drinkers, we owe this bit of information to a French inventor and scientist named Charles Cagniard de la Tour who, around 1840, discovered that yeast was the substance adding carbonation to beer. A few decades later Louis Pasteur released a book that studied fermentation and beer and, of course, pasteurization methods.

Once upon a time, brewers let the yeast do all the carbonating work. This is not true anymore, as we'll discuss in a moment. However, small-batch Belgian beers still rely on natural yeast-driven carbonation, and so do a few other styles that are historical in nature and don't need the extra push. When done properly, naturally carbonated beer can produce just as much CO_2 as a force-carbonated beer. There's also the perception that naturally carbonated beers have a creamier mouthfeel.

When it comes to force-carbonation technology—the process used by most brewers today—it helps to imagine an at-home

SodaStream. Specialty equipment allows brewers to regulate the CO_2 intake and make sure the gas properly dissolves into the beer. CO_2 works best when the beer is below 60 degrees Fahrenheit. This can be accomplished with a CO_2 tank and regulator: the carbon dioxide is slowly introduced into a keg, allowing the liquid to absorb the gas over time, achieving perfect serving carbonation. Carbonating slowly—typically, over five to seven days—both gives the beer a chance to age and increases the likelihood that the beer will not overcarbonate. This is important, because an overcarbonated beer will foam like crazy when tapped. It may even rupture tanks.

Another way to force CO_2 into beer is by using a carbonation, or diffusion, stone. With its long shape and fat head, this device vaguely resembles a spark plug. Popular with brewpubs, it is attached to tubing connected to the CO_2 tank and lowered into the fermentation tank prior to serving. CO_2 is forced through the stone (actually made of stainless steel) with increasing pressure, creating tiny bubbles that are immediately absorbed into the water. A carbonation stone is incredibly porous and should never be touched by hand prior to use, as skin oils can clog it and cause less than perfect results.

If you drink enough, especially if you're a regular imbiber of the same beer over time, you'll be able to tell just by sight, well before your first sip, if it has been carbonated correctly. Like Goldilocks, you don't want it too flat or too robust. It needs to be just right.

What is "just right" will of course depend on the style. When most American drinkers think of natural carbonation in a beer, they're usually thinking of cask ale. They will cite "lack of carbonation," or how the beer appears to be flat. Cask ale is traditionally served through a hand-pull tap or directly from a cask, where it has been carefully aged and allowed to develop without the aid of machines or modern brewing technology.

If kept correctly, a cask ale offers a subtle, natural carbonation which leaves a gentle gas prickle on the tongue, as opposed to the harsher sensation delivered by a CO_2-pushed beer. Cask ale, also known as traditional ale or real ale, is very much a living product. After the ale is brewed it is transferred to metal (or, less frequently, wooden) casks and then pitched with a fresh dose of yeast. This process is known as secondary fermentation and is what creates both the rounded flavors of the ale and the additional natural carbonation as the yeast chews on the beer's sugars a second time. The casks are stored in the pub cellar, where it is the job of the cellar staff to maintain the proper storage temperature—usually 55 to 57 degrees—and to ensure that the beer ages to the brewer's specifications. In some cases a cask needs only a few days to reach its desired level of carbonation. In other cases it can take several weeks or longer to achieve perfection. The importance of the cellar staff must not be understated; the profession is vital to ensuring that customers get the freshest and best-tasting beer possible.

"The implication that cask beer is meant to be flat is incorrect and is damaging to the growth of beer as such in its most natural form of dispense," noted cask ale expert Alex Hall once told me. Hall has joined others in saying that the aggressive, extraneous gas in a CO_2 beer can take away from the full mouthfeel. Linked with this issue is the serving temperature of modern kegged beer. For cask ale to achieve the fullest flavor spectrum possible, you want to serve and drink it at cellar temperature. If you cool it down past that, a lot of flavor nuance is lost, and the carbonation, already gentle, is reduced. Still, even when cask ale at its optimum state, the majority of Americans, because we are used to ice-cold, full-of-foam pints, will think of cask ale as "warm and flat." It's time to look beyond that, because any beer that is truly warm and flat would be undrinkable, and cask ale in proper condition with a gentle, naturally produced carbonation is many miles away from that description.

Traditional cask ale is under assault, especially in the United Kingdom, where the number of pubs serving the once-dominant style are adding more and more kegged beer. New breweries tend to favor modern recipes and serving methods because that's what their young customer base demands. While it's possible to find proper cask ale in cities like London, the style is less common than it once was and has been on a decline for the last several decades.

At least one group is dedicated to preserving this history and changing the associations that many have with cask ales. The Campaign for Real Ale (CAMRA) was launched in 1971 by a devoted group of beer drinkers who rose from their bar stools to found an organization that would protect the heritage of cask ale, promote its consumption, and educate new generations on its importance.

Despite its being synonymous with British beer, cask ale had been in decline since the 1960s. Lagers and other beers served from pressurized kegs were gaining in popularity, bottle sales were up, and the management, space, and upkeep required to serve cask ale were becoming bothersome to some pub owners. It seemed like cask ale would soon be extinct. However, today CAMRA has nearly two hundred thousand members and more than two hundred branches, and is the largest single consumer group representing ale and pubs. They regularly lobby the British government on ale-themed issues like taxes, pub ownership, and below-cost alcohol sales. They are an undeniable force in politics. In part because of their influence, it's unlikely we'll see cask ale disappear in our time.

Too often, CAMRA members are described as men with beards who wear sandals. That's not always a compliment, and nor is it true any longer as younger and more savvy members, including a strong number of women, have joined their ranks. The organization's growth offers more proof that people care about what's in their (collective) glass, where it comes from, and how it is cared for.

Aside from being a voice for cask ale (and more recently hard apple cider and pear cider), CAMRA hosts a number of beer events, including the Great British Beer Festival, held every August. The organization also releases a number of beer-themed publications. Among them is the *Good Beer Guide,* which rates real-ale pubs throughout the United Kingdom, awarding the best of the best. One of CAMRA's latest missions involves getting pubs to sell more locally brewed ale. Through their LocAle initiative, they shed light on the positive aspects of drinking beer made close to home, and the good it does for both customers and businesses. They realize that people want to drink fresh beer, to know where it's sourced, to support the environment, and to help local businesses.

Thanks in part to the work of organizations like CAMRA, cask ale is everywhere these days, even here in the States. If you've attended a ceremonial tapping you might have witnessed the opening of a firkin (a nine-gallon cask): a spile with a hand-turned spigot is hammered into the front of the keg, allowing for manual pouring. Cask ale can also be attached to a hand pump (or beer engine). You've no doubt seen such a cask sitting on a bar, covered in an insulating "blanket": it's usually filled with ice, like a cool pack, to keep the beer from getting too warm, and needs to be changed out every few hours.

Still, because we lack the strong cask ale tradition of the UK, American cask ale tends to be problematic. There are thousands of breweries in this country, and a good number of them claim to have a cask ale program, although very few do it well or even come close to traditional methods. There are some exceptions: Heavy Seas in

Baltimore has a robust cask program that stays true to traditional methods, meaning every time you order a cask ale it should be as close to an authentic British experience as if you'd actually traveled across the pond. Still, such breweries are in the minority.

Many other breweries treat cask-conditioned ale as a playground, often with middling results. They may take a regular recipe, say an IPA, and transfer it to a cask without prepping it for secondary fermentation. Then they'll dose it with something like carrots, ginger, or another complementary flavor. They'll call it "rare" or a "special edition," and while it might have a nice flavor, it is far from traditional, and it fails to demonstrate the great skill required to brew a cask ale. Most American bars lack properly trained cellar staff; too often the casks are treated the same as carbonated kegs, resulting in a beer that tastes extremely off.

If you're into the idea of a coffee-and-chili-pepper-infused porter, or a sesame-lime pale ale on cask, especially if you're already a fan of the base beer—go for it. Live a little and enjoy the experience. Just please don't call it traditional cask ale.

I don't know if we'll ever see a true resurgence of cask ale here in America. It might just be too quaint for the modern drinker's palate. That said, a soft, slightly malty, and lightly floral-hopped English mild or bitter can be a lovely drinking experience. It's a throwback to simpler times, and a pint of cask ale almost demands that the drinker slow down and savor the experience. It rarely gets the marquee treatment of the latest trendy beer, such as an ale hopped with the "it" variety of the moment. Cask is often an afterthought, and its creation largely relies on the few dedicated beer souls who appreciate a different kind of craftsmanship and the pleasure of a quiet pint.

If cask ale is relatively calm, the opposite would have to be nitro. There is something poetic, almost romantic, about a freshly poured pint of dark ale from a nitro tap. The cascading effect is mesmerizing; the waterfall of tiny bubbles slowly yielding to a dark brew with a fluffy, white head thick enough to float a bottle cap.

At this point it's okay to ask, "What is nitro?" The term refers to nitrogen (N_2), the type of gas used to add bubbles to the beer, a process called *nitrogenation*. (The term *carbonation,* of course, refers to the fact that *carbon* dioxide is the gas being used.) Nitrogenation creates a creamier beer compared to its lively, prickly CO_2 counterparts. In fact, the gas in a typical nitrogenated beer contains about 70 percent nitrogen and 30 percent carbon dioxide. Nitrogen is largely insoluble in liquid, which is what contributes to the thick mouthfeel. This characteristic is helped by a special piece of tap equipment known as a restrictor plate that forces the beer through tiny holes before it lands in the glass, causing the rising effect that is topped with the head. And it's really only the bubbles on the sides of the glass that fall. Inside they are actually rising, as is typically seen with any poured carbonated beverage.

The nitro style is normally associated with Guinness because the brewery invented it. This makes sense because the brewery has spent countless dollars on development, advertisements, and product placements and has installed branded nitro taps at bars across the country. Bars that have these taps, which are maintained by the brewery or its local distributors, must commit to pouring Guinness and only Guinness from those particular faucets. Realizing the appeal that nitro beers have for customers, and always looking to expand market share, smaller breweries have jumped on the nitro express in recent years, making beers specifically for the gas mixture and installing specialty taps in their tasting rooms. Bars are following suit and adding independent (nonbranded) nitro taps and a rotating selection of beers to go with them. It's hard to find a smaller brewer that hasn't at least dabbled in nitro, be it on draft, in bottles, or in cans. This includes Samuel Adams (Boston), Sierra Nevada Brewing Company (California), Sixpoint Brewery (Brooklyn), Sly Fox Brewing (Pennsylvania), Yards Brewing Company (Pennsylvania), and more.

One reason why smaller brewers have embraced the method is because nitro adds a level of complexity to beer. And pouring it is compelling to watch. Jim Koch of Samuel Adams calls the nitro pour "the great theater." As I mentioned in the last chapter, Guinness's best sales gimmick is when bar patrons watch the drama of a pint of Guinness being poured for another customer. The desire to order one of your own becomes very real. I want one right now.

Since there's no "nitro beer guild" or association, there is no official tally on the number of breweries that produce nitrogenated beers. However, because of the complex science (and the bit of secrecy) that goes into canned or bottled offerings, the smaller the brewery, the more likely they are to stick with draft. When it comes to cans and bottles, nitrogenated beers are equipped with a widget: a small plastic ball filled with nitrogen that releases when the bottle or can is opened, infusing the liquid with the gas. Most packages have instructions for how to properly pour these beers, but the general rule of thumb is to open and then pour hard into a glass, causing a great rush of turbulent beer that will quickly settle into that familiar rising look. Afterward, if you give the bottle or can a shake you'll hear the widget rattling around (unless the brewery has affixed it to the bottom of the package).

Left Hand Brewing Company in Longmont, Colorado, offers a milk stout as part of its core lineup. Once the standard CO_2 version of the stout hit shelves more than a decade ago, the brewery wanted to come up with their own proprietary technology for nitrogenating bottles of the beer. After a lot of trial and error, employees Jake

Kolakowski and Mark Sample figured it out. The brewery first released a bottled nitro version of its milk stout to the public at the 2011 Great American Beer Festival in Denver, where it was an instant hit and quickly became the brewery's flagship brand.

Many breweries now offer the same beer on both kinds of tap—forced CO_2 and nitro—side by side. The difference between the two styles is vast, with the nitro beer taking on a creamy texture and more evenly distributed flavors, while the CO_2 beer is more aggressive on the tongue and aroma forward. If you can find the same beer available in the two separate pours, doing a side-by-side comparison is quite a bit of fun.

Sharp-eyed drinkers will notice that most, though not all, beers served on nitro tend to be styles that are traditionally more malt heavy than hop forward. Thus, more porters and stouts than India pale ales are typically found on nitro. This is because malt plays better with nitro. That's not to say brewers don't experiment, but most have found that combining an IPA with nitro strips away a lot of the beer's essential hop oils, aromas, and flavors, turning what could have been a vibrant double IPA into more of a weak pale ale. That doesn't mean nitro IPAs are going away. Because IPA is the best-selling craft-beer category, brewers are continually trying to find new ways to get those three letters in front of drinkers. Guinness has a nitro IPA, Samuel Adams released one nationally (then took it off the market following poor sales), and brewpubs around the country regularly experiment with it. The truth of the style is revealed in the reviews, where nitro IPAs consistently receive a thumbs down for aroma and taste. Stouts, Scotch ales, barleywines, wheatwines, even hefeweizens all do well on nitro, though, bringing new depth to long-familiar styles.

Ultimately, the beers that are poured on nitro have been made to be served that way. Hooking a regular carbonated keg of, say,

American light lager to a nitro tap will do the beer no favors. Because of the way draft systems work, that nitrogen and CO_2 mixture will not serve most styles well. If a bar unwisely chooses to use the nitro system on ill-suited lagers and ales, they will quickly find themselves with kegs full of undercarbonated beer that will go flat within days.

WHETHER THE CARBONATION SYSTEM IS CASK, FORCED CO_2, OR nitrogen, the goal is always a properly carbonated beer. A significant amount of beer is sold on draft, so breweries, bar owners, and draft-system manufacturers make a concerted effort to ensure that each pint has the desired level of carbonation when it is placed before a customer. Too much carbonation can be unpleasant. But the real dread, of course, is a flat beer. Without the appropriate pop and fizz, a beer can be practically undrinkable.

Remember when we talked about aroma being one of the most important parts of tasting beer? Carbonation helps lift aroma from the depths of the beer to the surface and into our nostrils. There, the scents of hops, malt, and anything else the brewer chose to use in the batch comes to life and gives us a sense of what's to come.

You might be wondering, as I once did, does CO_2 itself have a taste?

"We tend to think of beer as being basically hops on one hand and malt on the other hand. It is more complicated than that. Carbonation puts an acidic quality into beer," says Randy Mosher, a Chicago-based author and beer consultant. "Beer is mildly acidic anyway, but carbonic acid adds more."

To explore the topic a bit deeper, let's use the American light lager as an example. Carbonation is the most intense sensory experience the style offers. There are almost no hops, there is not a great deal of malt, and even the yeast is weak; what we're left with is carbonation, and quite a bit of it. That level of carbonation suits the style, but in other styles a mild tingle of carbonation is enough to stimulate the taste buds and give us a chance to feel bubbles on the tongue while getting a kick of hops and the tang of malt. Carbonation also plays a crucial role in a beer's mouthfeel, and in some cases that's the most interesting thing a brew has going for it.

We should also pay attention to carbonation for potential warning signs that a bar experience might be less than optimal. Two recurring problems can crop up, thanks to issues with hygiene (unintentional) and advertising (intentional): carbonation that sticks to the inside of the glass, forming a curtain-like façade; and a frosty cold glass, served right from the freezer with ice caked on the rim and sides.

Every time you see a bubble clinging to the interior of a glass, it means there is something between your beer and the glass itself—something that should not be there. It may be anything from soap, to lipstick on the rim, to food deposited there by the dishwasher, to sanitizer from a rinse. At worst, it's an age-intensified problem caused by all the above. It means your glass is dirty.

Regardless of glass type or shape, or even what's inside, there is never—NEVER—an excuse for a beverage to be served in a dirty glass. And to that you'll say, "Of course, I know that." But it happens more often than you'd think, and a good number of people, even when it's right in front of them, don't realize they're drinking from a dirty glass.

Dirty glassware can end up in front of a customer even at establishments that take their beer selection seriously. Not too long ago I was at a bar in my neighborhood, a place not in my usual rotation but with an expansive tap list and many televisions, making it an ideal spot to catch a game or three. That day, I was waiting for a friend to pop out of the train station. I sat at the empty bar and ordered a lager. When the shaker pint of beer arrived, I had a tough time seeing into the straw-gold liquid because of the thick layer of carbonation lining the inside of the glass.

I struggled to comprehend the staggering level of filth that must have accumulated in the glass to render it nearly opaque. The bartender who served the dirty glass was talking to a coworker about a recent beer trade he'd just made, touting his own "serious beer geek cred." This was a man clearly in the know. When I caught his eye and told him that I needed to send the beer back, rather than acknowledging the dirty glass, he simply assumed that I was inexperienced in beer flavors and didn't like the taste of this "craft lager." It's not for everyone, he told me, but he was still going to charge for the pint. With my Irish temperature rising, I informed him that it wasn't the taste; it was his serving an otherwise excellent beer in a dirty glass. He sheepishly took back the glass and tried to offer me a clean one, but my stomach had turned and I simply left. Months later, I know that was an extreme reaction on my part. But life is too short and beer is too good to accept a nasty glass.

This is a topic that I'm especially passionate about. If I'm going to spend money on a beer, I expect that regardless of what I order it will be served in sanitary conditions. However, because carbonation is a natural feature of beer, some people think that stuck bubbles are either part of the experience or just nice to look at. I've even seen stuck carbonation used in beer ads, indicating that sometimes beer makers themselves don't think of this as an issue. Not long ago, I watched a documentary that focused on two small breweries and the various challenges they faced. Video chronicling the opening night of one brewery had a shot of a logo pint glass filled with beer and—you guessed it—carbonation stuck to the glass's interior.

We've come a long way in terms of getting quality liquid choices on tap, but we still have a ways to go in other areas. If you were at a restaurant and your fork showed specks of gunk from its last use, you'd ask for a clean fork, right? We, as educated and passionate beer drinkers, need to speak up as well. If you are served a dirty glass, you don't need to make a scene, but you should take the opportunity to point it out and explain why it turns you off. I'm a fan of doing this in person, politely, rather than taking to social media or review sites like Yelp. When I've done so, I've often encountered resistance, a natural defense to embarrassment, but generally people take the suggestion to heart. Hopefully, the bar will then reevaluate its process. Thankfully websites and training programs are available, like the earlier-mentioned Cicerone Certification Program, to teach breweries, bars, and restaurants about proper service and cleanliness. If I identify a bar as a repeat offender, I don't mind informing a brewery that a trusted account is letting down their beer. And when all else fails, I vote with my feet and wallet.

Let's return to the other problem common at bars across the country: the ice-cold glass. Envision a shaker pint or handled mug pulled directly from the freezer; when it is filled with beer, a large amount of frothy foam flops over the top. That extra foam results when the ice on the glassware interacts with the beer, rapidly cooling the already cold product and causing it to break apart, creating foam. This can mean the beer you taste doesn't resemble the brewer's intent.

Furthermore, if you store the glassware in a freezer with other items, as the ice crystalizes it has the potential to take on other flavors and aromas that might be kicking around in the freezer. You don't eat the crystals that form on your food as a result of freezer burn, right? Why would you do it with your beer glass?

Frozen glassware also stunts the flavor of a beer. While this isn't a big deal when drinking a beer like Bud Light, which doesn't offer a ton of flavors and actually goes down a little better when it's dancing right on the edge of freezing, it can impact other styles, like a tropical stout or barleywine. For these beers, the full range of flavors and nuances are best exposed when they're served at around 55 degrees, or cellar temperature.

This problem has an easy fix: if you notice beer mugs or glasses at your local bar being stored in the freezer, just ask for a nonfrozen glass. There are usually some available, and the server likely won't care either way.

Chilled glassware, rather than frozen, presents a different story. Some places refrigerate all glasses in advance of service. Again, depending on what else is in the cooler where the glasses are stored, you run the risk of them taking on foreign flavors. Although a chilled glass doesn't pose the potentially dramatic issues that a frozen glass does, it simply isn't necessary, especially if the beer has been stored cold and is served at its proper temperature. Chilled glasses are relatively harmless, but ultimately unnecessary.

If you are caught by surprise and served an icy glass or one that feels excessively cool, you have the tools—your hands—to get things back on track. Simply cup your hands tightly around the glass to transfer your body heat to it. It'll warm up faster than you might think (although you'll likely have to give your palms a break every little bit).

What I'm saying here might be confusing to you if you're a close observer of television commercials featuring certain beer brands. With scenes from the snowcapped Rocky Mountains, Coors talks of its "ice-cold" processes, from fermentation to bottling to your refrigerator. Remember, we learned that yeast, and especially lager yeast, likes it cold, so the Coors brewery is doing just what it's supposed to during the brewing process. But the ad wants you to think of the beer as refreshing, so it hammers home its frosty point. This is nothing unique to the brewery; it's just how lagers are made.

The one circumstance under which ice is welcome in beer is in the case of an eisbock. Although some brewers now use freezers to make eisbocks, traditionally they are made in winter, when the temperature stays below freezing for days at a time. After having made a traditional bock beer—usually around 6 percent ABV—the brewer places the beer in the subfreezing environment until the water separates from the alcohol. The H_2O is then removed, and the concentrated beer is ready to serve. The finished product is usually double the strength of a traditional beer, has taken on richer, usually fruitier and sugary caramel flavors, and is best enjoyed in small servings. It's a labor-intensive beer and not made very often, so if you come across one on offer, especially on a January or February day when you need something to take the chill out of your bones, don't pass it up.

This seems a good time to address a fairly common question: if we can put ice in soda, seltzer, cocktails, and other drinks, how come we can't put ice cubes into a glass of beer? First of all, you *can* put ice in beer, but you really shouldn't. Ideally your beer should be served to you at the correct temperature, meaning it's already cold and refreshing. If you feel like you need to add ice, your beer was served too warm. Further, ice dilutes the beer as it melts, messing up the intended ratio of ingredients. The only exception might be for shandies or radlers, summer beverages usually consisting of half juice and half beer. When the temperature is hovering around triple digits and we want to stay as cool as possible, one of these "beer cocktails" is fantastically refreshing with a couple of ice cubes in it.

Subjecting a beer to excessive heat should also be avoided. Brewers have spent a long time running trials to determine the perfect temperature for beer service. Exposing a keg to warm or rising temperatures (if you've taken it out of cold storage while serving) can cause excessive foaming as the CO_2 begins to separate from the liquid. Some yeast phenolics, or aromas, can begin to break down and change for the negative when introduced to the stress of heat.

Hopefully you won't encounter a place that keeps their glasses in the oven, but you're likely to encounter a hot-to-the-touch glass when it's taken right out of the dishwasher and has just been blasted with steam heat. If you see a glass come out of the machine this way and it's destined to be yours shortly, maybe hold off on that next round for a few minutes (the glass will cool rapidly) and use it as an opportunity to hydrate with water between rounds.

I can think of an exception to the rule about avoiding beer that's been subjected to heat (and it's the only exception I've ever encountered): August Schell Brewing hosts an annual Bock Festival in mid-February on the brewery's forested land in New Ulm, Minnesota, about two hours southwest of Minneapolis, People come from around the country to celebrate bock, a strong lager. They stand outside in temperatures that can reach –20 degrees Fahrenheit. To help keep folks at least a little warm, bonfires are lit around the festival grounds. Upon request, tenders of the fires will pull a red-hot poker from the flames and dunk it into your beer, causing the beer to foam terrifically and raising its temperature a few degrees to help you get it down a little easier. As a bonus, some of the residual sugars in the beer caramelize in the glass, creating a toasted-marshmallow effect. Now that's the right experience at the right place. Don't try it at home.

SIX

SHADOWS IN BEER

THE BEER INDUSTRY HAS CHANGED SIGNIFICANTLY OVER THE last two decades as more breweries have opened. The community of beer lovers has grown by leaps and bounds. Some high-end restaurants now carry impressive beer lists, better brews are served at ballparks, and several television shows on major networks have chronicled brewery life and the brewing process. Whereas in the past it was only niche publications and websites that chronicled daily beer life, the mainstream media now regularly covers the beer industry from business to culture. It's not uncommon to see beer reviews pop up in newspapers, or segments about seasonal beers on television. So many breweries offering so many choices has made the modern beer phenomenon almost impossible for society to ignore.

Yet for all that good, there is still work to be done if more newcomers are going to join the fold. I'm not talking about finding flavors that appeal to more people, but cultivating an environment and an attitude that are inclusive to everyone, not just to some. The beer industry is currently dominated by men. Specifically white men. I know this because I am a white, middle-aged man who covers the industry. When I'm at a beer event, conference, or other industry gathering, the room looks a lot like me (except I don't have a beard).

Breweries are traditionally staffed by men, so there should be little surprise, sadly, that now and again there appear beer names, labels, events, and work environments that demean women.

I'm talking about beer names like Panty Peeler. Phat Bottom. As equality fights rage on a number of different fronts, breweries and beers still exist that are well behind the times. Look, I'm not naïve enough to think that sex doesn't sell. Beer has always used women to promote products. In the 1970s, '80s, and even early '90s, bikini-clad models appeared on billboards and in magazine ads to promote light lagers. Going back further, in the '40s and '50s beer advertisements sometimes centered around "good housewives" who gave their husbands what they really wanted: "their favorite beer." In general, however, those ads and labels weren't crude. They were sexist, a rather sad sign of their times, but not crude. Unfortunately, in more recent times, some beer labels are actually crude rather than clever. The Brown Note, a brown ale from Against the Grain in Louisville, Kentucky, comes to mind. Not only does the cartoon label depict a soiled pair of underpants, but the brewery claims it's "so good it WIlL make you shit yourself!!!!" The capitalization and exclamation points are theirs. Sigh.

At the annual Craft Brewers Conference in 2016, the Brewers Association, a membership organization for "small and independent" brewers, held a press conference in which leaders of the group rolled out good news about growth and new initiatives, and also issued some warnings about what "big beer" could do to the industry and consumers. During questioning by members of the media, the writer (and my friend) Bryan Roth posed a question about diversity. What is the BA doing, he asked, to make small brewers more aware and supportive of "inclusivity and diversity, whether that's race, ethnicity, or gender identity?"

What followed were a number of vague statements from BA employees that ranged in topic from the economic to the geographical, to being celebratory of the industry, with a nod toward needed progress. (Since then, the group has stated publicly that it intends to research gender and race in the industry to see how it can do better in regard to fairness and attention to the issues, and it has already put some initiatives into action. It has a panel of industry members who look out for such things, and announced in 2017 that should any beer with a distasteful name win an award at the Great American Beer Festival, the name wouldn't be read aloud onstage at the awards ceremony. The group also named its first diversity ambassador, J. Nikol Jackson-Beckham.)

Inclusion and equality in the beer industry don't get discussed often enough in open forums. These are issues I noticed early on in covering the industry, but I didn't quite know how to address them—specifically, the way some breweries view and depict women in beer names and on labels. They might take a beer style, like a blonde or an amber, and sexualize it somehow (usually in a reference to a woman's measurements). This isn't a new issue. It's been simmering below the surface for quite some time, with only the occasional bubble-up. Some drinkers take to the internet to express outrage; others shrug it off. Still others criticize the critics. However, as the national political conversation centers on how women are treated in society these days—to say nothing of the xenophobia, racism, and homophobia permeating so much of everyday American life—it would be wrong not to stand up and join a call to action.

How, in an age of progress, technology, and enlightenment, are breweries still releasing and producing beers with demeaning names like Once You Go Black or Thong Remover? Or the crowdsourced name solicited by MobCraft Beer: Date Grape? To be fair, when the potential offensiveness of the name was pointed out to them, the Milwaukee-based brewer immediately pulled it, apologized, and promised to put editorial checks and balances in place to avoid future embarrassment and hurt.

How do you know when a label has crossed the line? Well, it's a lot like pornography. You know it when you see it. Sex can be celebrated, the human form can be admired, and yes, sex will always sell.

For me, the tipping point came a few months before my daughter's birth. My wife and I took a vacation to Alaska and, after a ten-hour plane ride, we picked up a rental car and headed to a local brewery where I could get a pint and April could get dinner. We walked in to find that the flagship beer at this particular brewery was a Belgian-style tripel called Panty Peeler. At 8.5 percent ABV, it's not a subtle beer. According to the brewery it was originally called Extreme Polar White Bier, but it got the underwear-removing nickname and eventually took on an official label change. The connotations are obvious, and knowing that I would soon have a daughter in a world where this kind of mindset was condoned was honestly upsetting. The brewery, for its part, claims that the beer is a celebration of a woman's independence. (It's worth noting that the brewery is owned by women.) But it bothered me anyway. Being acutely aware of my cultural and racial background, and of the sensitivities that can be sparked by speaking up, I was hesitant about my next step.

This is still a male-dominated industry, in terms of both its employees and its customers. In some ways this fact has allowed participants to operate in a vacuum. Outdated thinking, unchecked juvenile behavior, and simply a lack of balance have, in some circles, allowed the continuation of the jocular attitude that women are somehow beneath men or are simply objects. These attitudes should have been eradicated a long time ago.

At the time, I was the editor of *All About Beer Magazine*, and editorially we decided to take a stand. The magazine, I rationalized with the staff, existed to cover the beer industry in all its forms. We had a social responsibility to stand up against anything that demeaned our fellow citizens, regardless of gender, religion, sexual orientation, or race. We wanted to do our part so that the next generation of beer drinkers could focus on the fun, the flavorful, and the future, without getting bogged down in shortsighted intolerance. And so we decided that beers whose names or labels demean women or promote rape culture would not be reviewed or promoted in the magazine or on its website.

That stance wasn't radically new for us. For the most part, we already avoided covering or reviewing beers with names that fell into poor taste—unless the beers were otherwise newsworthy. We saw no benefit in rewarding juvenile behavior, and honestly the crude names almost always existed to distract the drinker from the poor quality of the beer. I wrote a signed editorial and closed it with: "Demeaning or objectifying women has no place in society or on beer labels."

The response was instant and wildly varied. Women and enlightened people in the beer industry praised the sentiment and shared it on social media. Many correctly pointed out that they had been banging this drum for years, but that it took a man with some editorial clout to carry the message to consumers. On the other side of the coin, folks in conservative media attacked the magazine, me personally (I was dubbed "chief of the morality police" by some right-wing blogger), and the staff at large for trying to tamp down on breweries' free-speech rights. Or they just accused us of not having a sense of humor.

My personal Facebook page was hit by strangers criticizing me, including one guy—under an assumed name—who said I'd probably like a beer called Cocksucker, because that's what I was. The owner of a bar in my city wrote me an email that said, "That might be the worst I've ever seen, man. C'mon, you're better than that." I didn't much mind that one, though. I had already stopped patronizing his bar because he served good beer in dirty glasses.

Conditioned by my years of being a straight news reporter, I usually stayed out of the fray and almost never offered personal opinion in print. By the time we published the editorial I had taken a few stands on other issues, but nothing quite like that third rail. I was surprised by the blowback. But more importantly, I understood in a new way that the criticisms and name-calling being hurled at me were nothing new for the women working in or covering the beer industry who've talked about and written about and fought back against these issues for years. In that light, the experience was humbling.

I do think that brewers go with these names simply because they don't stop to think about the ramifications or the larger social context. When all they see at work are dudes who look just like them, it's hard to break out of that bubble. At that point, they should step into the taproom. Because although the beer industry itself is male-dominated, there are an increasing number of women and minority beer drinkers. If breweries are the new pubs—and they are—looking around a taproom reveals a city's or town's true diversity. Women and men, young and old, from all walks of life gather to drink at a brewery, and you can be sure that exclusionary beer names, or ones that treat gender, sexuality, or assault as a joke, will either have to be changed, or the breweries that devised them will have to close their doors. Eventually—and especially in the current hyperaware social climate—consumers will vote with their dollars.

There are far fewer insulting beer names than ones that are actually funny, informative, or simply appropriate. But the offensive ones that do exist highlight how pervasive the problem is in just about every facet of life. No matter how you're drinking, or where you're drinking, beer is a social beverage that doesn't know gender, sexuality, race, creed, or political opinion. The act of getting together should be an inclusive one, while still promoting the enjoyment of beer. Beer is what we're supposed to drink while solving the world's problems (remember the Founding Fathers?); it's not meant to be a signifier of cultural problems. All of us who participate in beer culture have a duty to make sure that a beer's label, the ethos behind its creation, and the intent of the brewery adheres to a progressive social standard—one that lifts up everyone and never puts down anyone.

WHILE THE SEXUALIZATION OF WOMEN BY THE BEER INDUSTRY HAS been an issue for decades, the truth is, even eighty-plus years after the repeal of Prohibition, this country has other concerns surrounding alcohol. Some are based on puritanical or moral religious standards, others on health concerns or social norms. One of these is the debate about the legal drinking age.

We've all heard the argument: if we can send soldiers overseas to fight at the age of eighteen, if you can legally buy cigarettes, play the lotto, and vote, why can't you also buy a beer?

The answer is more rational than you might think: traffic studies clearly show that the number of alcohol-related fatalities decreased in young people when the drinking age was raised, according to the National Institutes of Health. There's still a large percentage of the population who remember when you could legally drink at eighteen. The National Minimum Drinking Age Act of 1984 bumped the age in all fifty states up to twenty-one. That step, like so many other restrictions on alcohol manufacturing and consumption, was taken with safety in mind, specifically to reduce occurrences of driving under the influence and of binge drinking.

Please don't get me wrong: no one should get behind the wheel when impaired, be it from beer, wine, spirits, pills, marijuana, or any other substance that can alter the human condition. Bad things can and will happen, and such accidents are easily avoidable. But additional laws and restrictions don't stop binge drinking. Turning twenty-one is a rite of passage now, and it's common for young adults to spend that birthday drinking in excess because they can. That's not healthy behavior either.

What I believe we need is better education about how to manage alcohol, and we need to be a bit more liberal in the way we approach alcohol with younger teenagers. I'm not saying we should give minors full servings of beer or glasses of wine, but allowing them to try alcohol at a young age, in a responsible way, shows them that consumption can be a healthy part of a dining or social experience. It can lead to greater responsibility later in life and lead them to treat alcohol as something that shouldn't be abused but respected. By making alcohol an approachable and normal part of life, not hiding it away like the forbidden fruit of Eden, alcoholic beverages will become less of a thing to sneak behind your parents' back and more of a shared experience. That's one of the things I like about modern breweries—they're gathering spots for families. If a kid is brought up seeing responsible drinking behavior, associating it with a special occasion or a fun experience, they are more likely to respect the responsibility required of them when they reach drinking age.

I'D LIKE TO TOUCH BRIEFLY ON A TOPIC THAT'S PERHAPS EVEN MORE controversial: an occasional glass of beer consumed while pregnant. You should definitely follow the advice of your doctor; my wife's

OB-GYN gave her the all-clear for the sporadic adult beverage after the first trimester. Excessive drinking during pregnancy can certainly harm a child and should be avoided. But even drinking in moderation while pregnant remains a serious taboo, though it isn't illegal. Any establishment where alcohol is served will post signs (usually an outline of a pregnant woman holding a glass with an aggressively large red slash through the image) warning about the health risks.

Let's remember that for centuries, before methods were developed to sanitize water, low-alcohol beer was the preferred source of potable water for children, the elderly, and pregnant women. In Ireland, stout has been called "mother's milk," and certain nutrients in malt and yeast can actually be good for the body. Still, the stigma is real, and a visibly pregnant woman at a bar or restaurant imbibing an adult beverage will likely be met with looks of scorn or outright judgment. The modern beer movement has eased some of that, because breweries are, by and large, "no judgment zones" for pregnant women. Women that I've spoken with about the phenomenon, including my wife, say that it's mostly because the staff and management at today's breweries (as well as the clientele) are forward thinking, and are more apt to leave women to their own decisions. A beer at a brewery is an indulgence a pregnant woman can usually enjoy without being on the receiving end of preachy comments or stink-eye looks. Some women choose to abstain during pregnancy, and others practice moderation. If you're in the latter camp, a brewery is usually a fine place for a break.

IT'S WHEN WE MOVE BEYOND MODERATION, FOR PREGNANT WOMEN and everyone else, that things start to be dicey. I remember walking around the Great American Beer Festival a few years ago and seeing a guy in a T-shirt that read, "Craft beer isn't alcoholism, it's a hobby." Well, there's a thin line between the two, and just because you're not downing a six-pack each night doesn't mean that things are okay or that your intake is healthy. Alcoholism isn't easily discussed or addressed. Although some people might have the wrong idea that alcoholism is confined to the destitute or depressed, in fact, it affects 12.7 percent of the US population, according to a 2017 study in *JAMA Psychology*. That means it impacts family, friends, and neighbors, including professionals and folks from all walks of life. Because beer drinkers can be such a tight-knit community, we have a responsibility to know when our drinking companions have had enough, and if we suspect someone might have a problem, to try to find ways to help them. The same is true with finding strength in ourselves to acknowledge if we have a problem and to seek help.

Since we're talking about issues related to overconsumption, it's time to point out that the term *beer belly* exists for a reason. When it comes to calories and carbohydrates, beer is not water. It's referred to as drinkable bread! If you enjoy beer, you run the risk of needing to buy new pants. Yet, despite the climbing number of breweries, anecdotal evidence from beer festivals and brewing establishments seems to indicate that many drinkers work hard to keep their waistlines in check. Beer drinkers, long associated with the physique of the character Norm on the television show *Cheers*, or that of Homer Simpson, are more active than ever. This is thanks, in part, to the lifestyle activities—biking, skiing, disc golf, and running—that are regular parts of brewery employees' lives.

That said, the more I look around at beer festivals, the more I realize I'm in the minority now, that I'm less healthy than many others who attend these same events. While some beer lovers fall into the overweight category—myself included—beer has also become a part of social events that involve physical activity, where it's treated as a reward for hard exercise. Breweries host weekend-morning yoga classes before they open for regular business, or sponsor running or biking clubs. You can even participate in outside-the-brewery activities like the "beer mile": a mile-long fun run that rewards you with a full beer at each quarter-mile mark.

Still, beer is not necessarily healthy—no matter how much we want to believe the occasional studies on the local news that insist a few pints will help you live longer. The fact is, if you want to be a regular beer drinker and also be healthy, you need to find the right balance between activity and consumption. I've struggled with my weight since high school, but, spurred on by the example of my wife, coworkers, and friends, I've set out to lower the number on the scale. I'm walking every day, taking days off from imbibing and sampling (this job requires trying a lot of beer), and drinking a lot of water. I'm watching my diet, too. It's too easy to order the burger and fries—rather than a salad—after a few rounds. But because there's so much great beer out there and so many new places to explore, I want to be around long enough to experience the excellent things in life while being in good health.

As is the case with so many others who struggle with their weight, it took me a while to get to where I am today. I know I'm not alone, and for those of you in the same camp, there's no better time to make a positive change than right now. The challenge is this: let beer be only one part of a fulfilling lifestyle. If you need to shed a few pounds and want to feel better, plan physical activities around your nights out, bottle shares, and casual quaffs. If you need the extra motivation, get your buddies in on the act and do it together. Set goals, with special bottles being the reward. Together we can all be healthier and enjoy great beer, and we don't need to do it alone.

Nor do we need to compromise our fondness for top-notch beer. A few years back a book series came out that urged people to eat or drink "this" and not "that." The force behind these books, the editor of *Men's Health,* was on the *Today* show one morning and pulled out a Sierra Nevada Bigfoot barleywine, a 12-ounce bottle of thick, boozy goodness that's released each winter. This bottle, he said, aghast, is teeming with calories. Don't drink it; drink Michelob ULTRA instead. He honestly suggested a 64-calorie lager as a comparable trade for a rich barleywine!

In that moment he missed the point. You don't drink Mic ULTRA for flavor; you drink it for hydration, or, let's be honest, in large quantities to get drunk while hoping you won't expand your gut too much. Bigfoot is full of flavor and nuance, and drinking one at the end of a day won't set you back too much on the calorie-consumption limit so long as you've kept everything else in check. Recommending that a drinker sub a "diet" beer for a more robust one is just a silly argument, like recommending André Cold Duck in place of Krug because it's cheaper. You drink good Champagne when you want to splurge.

And lest you think it's just the larger breweries peddling the low-calorie beer, consider SeaQuench Ale from Dogfish Head, described by the brewery as a "4.9% ABV session sour brewed with lime juice, lime peel, black limes, and sea salt." The brewery's press release markets it as a "low-calorie indie craft beer loaded with flavor and light on the carbs." *Men's Health* called it the best low-calorie beer of 2017. Other breweries are following suit. Lakefront Brewery in Milwaukee, Wisconsin, released a 99-calorie green-tea-infused beer to the marketplace in 2018. "It's fairly difficult to craft a great-tasting, full-bodied beer that's this light," said Michael Stodola, Lakefront Brewery's brand manager, in a media release. "Our head brewer,

Luther Paul, and his team have nailed the liquid, the calories, and a beautiful green-tea flavor. We're using a great, citrus-forward yeast called 'Juicy,' Lemondrop hops, and a lot of not-so-cheap green and oolong tea. It's a high-quality, light craft." Expect to see more breweries adopt this language, exploring lower-calorie offerings to appeal to a certain segment of drinkers while pushing the flavor envelope.

Truth is, too much of any style can lead to the dreaded beer gut. One of the things I appreciate about beer today is that breweries are aware of this problem, and often are committed to helping consumers stay healthy by sponsoring the sorts of active-lifestyle events I mentioned earlier. Yes, part of the fun is being able to enjoy a beer after the exercise, but it's a pint that you've earned through hard work and in the company of like-minded people.

For all the good times we have with beer, it's important to remember that shadows lurk around the edges of our favorite product. If we beer drinkers equate the act of drinking with pleasure, we also have a duty to be responsible, to speak up against things that go against our ethical or moral code of conduct, to be understanding and welcoming of all who walk through the doors of a brewery, bar, or our home, and to offer a helping hand to friends and family who may appear to be struggling with excessive drinking.

Celebrate beer, yes. But make sure that what's happening around it is inclusive, responsible, and respectful. If it's not, speak up.

SEVEN

BEER AT HOME

As much as I enjoy drinking beer at a pub or a brewery, I also enjoy quiet nights at home, a beer in my favorite glass, catching up on recreational reading, or, on Sundays, having a pint while dinner cooks and football beams through the big screen. Beer at home also means that when friends come over I can dig into the stash of special bottles and cans I've collected while traveling, open them, and hang out with good people over (hopefully) good beers. And the ones that don't make the cut? I can drain pour them without guilt or outside judgment.

I know that this experience isn't unique. I don't need surveys, spreadsheets, or analytics to tell me that more people than ever are drinking better beer. All I need to do is stroll through my neighborhood on Thursday nights, when my neighbors put out their recycling for pickup. On those nights, alongside my trusty mutt, Pepper, I meander the streets of Jersey City, NJ, on our final walk of the day, taking my time to unwind and let the dog do her sniffy thing. I don't go digging through barrels; I just observe whatever is on top of the pile. There, among the crumpled Poland Spring water bottles, are empty bottles and cans from the likes of Maine Beer Company, Departed Soles Brewing, Two Roads Brewing, Boulevard Brewing, Industrial

Arts Brewing, and more. As the seasons change, so do the beers. Summer brews turn into bottles of pumpkin ale, followed by winter ale empties.

The first article I wrote on beer appeared in print in 2002. It was a relatively short piece on how the brewing scene in my home state of New Jersey had grown over the previous five years and was poised to grow even more. I had discovered beer through my job as a newspaper journalist. Traveling the country to report on stories of mayhem and everyday life, I sought out brewpubs in the evenings. There, I could find friendly people, knowledgeable staff, and a community spirit that made life on the road a little easier. Through those interactions, I became well versed in beer and the modern brewery scene.

At the time, I had no idea that beer writing would become my full-time career. I've been extremely lucky to chronicle the story of American beer for the last decade and a half. I have enjoyed the efforts of skilled brewers, met kindred drinking souls, and visited many states to see the business firsthand. It's an extraordinary privilege.

There are downsides, however. Covering the beer industry means living in a bubble. It's a realm where nearly everyone I interact with knows about hop varieties and geeks out over barrel aging. Where names of celebrity brewers are batted around casually, and plans for a drink are made not at the local bar, but at large events like SAVOR (a commercial event celebrating the pairing of beer and food) or the Craft Brewers Conference (an industry event). It's a world unto its own, and I often feel that it leaves customers or casual drinkers out of the equation. That element has always been particularly troubling to me, because as a journalist it's my job to educate, inform, and entertain the general public.

That's the reason I like having friends over, especially those who aren't around the beer world, and why I like casually eyeing the recycling bins at night. I'm able to see firsthand what's popular in my neighborhood and what people stock up on for parties, and occasionally I spot a bottle that is unfamiliar or is of a vintage that makes me think someone just celebrated a special occasion. These observations take me out of the bubble and remind me what the everyday consumer is interested in.

If you as a beer lover leave your house to drink, there's a good chance that you're at a brewery or a bar drinking draft beer. That's the smartest way to ensure you're getting quality beer. When that's not feasible, however, you're left choosing between bottles and cans. I'm often asked which is a better vehicle for beer, and the answer is, honestly, it's up to you. The fact is, if a brewer is putting a beer into a package and wants you to purchase and consume it, they are going to make sure—no matter the method—that the beer is at its optimal condition when it arrives in your hands. However, each of the three options (draft, bottle, or can) has its own appeal, benefits, and place.

"IT'S A TESTAMENT TO FORM AND FUNCTION THAT THE BEER BOTtle has endured so long and changed so little. As sleek as airplane wings and smooth as polished stone, these laborers of the beer industry sweat like construction workers, sing like wind chimes when clinked and sigh like happy men when relieved of pressure. They're an ever-present part of the beer experience."

That's the lede of a story I asked my friend Nate Schweber to write for *All About Beer Magazine* in 2015. In just a few sentences he perfectly summed up the often-overlooked workhorse of the industry and the romance it can bring to the drinking experience.

Bottles and beer go together like peanut butter and jelly. Likely first used in the sixteenth century, the bottle, much like the beer, has evolved. Once hand blown by glassmakers, today bottles are mass-produced by the thousands. Bottle glass is made from recycled glass, sand, soda ash, and limestone. The ingredients are mixed and melted at 2,850 degrees Fahrenheit until the blend forms a red-hot liquid. It's divided into smaller quantities, then fed into a mold and shaped into the desired bottle size. Beer bottles in the United States come in four main sizes: 12-ounce and 22-ounce bottles that are typically capped with a crown; or 330-ml and 750-ml bottles that look more like wine bottles and are topped with a cork and cage. There are other sizes, of course, such as the 7-ounce bottle (some call them pony bottles or nips) that Corona releases under the Coronita Extra label. Some brewers have even released 1.5-liter bottles (magnums).

There are only a handful of glass-bottle manufacturers in the country, and although they offer some choice in bottle size, color, and shape, in America we're likely to encounter only a few shapes. The first is the traditional 12-ounce longneck, the kind used by Budweiser. There's also the 12-ounce heritage bottle, used by Sierra Nevada, which is more rounded in the middle and has a shorter neck. Then there's the 11-ounce stubby bottle, which is almost all body and no neck. Red Stripe made this format famous, and Full Sail in Oregon uses it for their Session line of beers. And of course we can't forget the 40-ounce bottle, usually topped with a plastic screw cap. It traditionally holds malt liquor, a high-octane beer with the primary function of getting the imbiber drunk.

Brewers have done their best to personalize bottles. The signature of Samuel Adams appears in a raised form on each bottle from that brewery. New Belgium uses a raised collar where the neck meets the base to display its name. Budweiser has its trademarked eagle. Even with the label stripped off, there's no mistaking what is, or was, in that bottle.

Brown is the most common color for beer bottles. This is a conscious choice. Sunlight is a great enemy of beer, especially hoppy ones. When sunlight hits hops, the blue ultraviolet ray negatively interacts with the alpha acids in hops, creating that familiar skunky aroma. It is one of the most easily identifiable aromas in beer. Clear glass offers no protection, which is why you won't see pale ales or IPAs packaged in clear bottles. Green glass offers some protection, but still admits sunlight. Brown glass protects the contents the best, although even it doesn't do so absolutely. And black glass just isn't practical on a large scale due to manufacturing costs.

So why do some breweries still use clear and green glass? Some of it is marketing, and for a select few others it's by design. Clear glass is typically used for Mexican beers and other lagers with no discernable hop content. It conveys a cool, refreshing, sunlight-and-relaxation-in-a-glass vibe. Miller High Life, the Champagne of beers, as it is known, also comes in clear bottles, but that brewery uses hop extracts that aren't damaged by sunlight.

Now, think of a green-glass bottle. Ten to one you pictured Heineken, the lager from the Netherlands with the identifiable red star on its neck. Heineken is a juggernaut, the namesake beer from the second largest brewery in the world. Heineken, like other European imports, used green glass when their products first arrived on American shores in the mid- to late 1900s so they would stand out on shelves. Among rows of brown glass, the green was quickly identifiable and even had an air of European sophistication. It's also the reason that just about everyone in America thinks Heineken should taste a little lightstruck—that is, a little skunky. In Europe, when you find fresh Heineken on draft it tastes like a completely different beer: crisp with a kiss of noble-hop bitterness. Heineken used brown glass in its home market and throughout Europe until about ten years ago, when it switched primarily to green glass to capitalize on its global brand recognition. If skunky aromas were hurting sales and the brewery's bottom line, it's a good bet they would stop, but the continued use of the green bottle tells us that most people just don't mind. Heineken also cans its beers, and while the canned version tastes closer to the genuine article, most drinkers still perceive that old familiar lightstruck aroma.

It doesn't take long for sunlight to skunk your beer, and in most cases this will make it undesirable. If you've poured the beer into a clear glass, it can take less than a minute, depending on the hop content. After four minutes in direct sunlight, even if kept in a brown bottle, you're in skunk city and should just toss the beer and go open a fresh one in the shade. There is at least one style, the Belgian saison, in which very low levels of a lightstruck aroma are acceptable. These farmhouse ales typically use aged hops, in very small quantities. Traditionally they were packaged in green bottles, and in the case of the classic (and benchmark for the style) Saison Dupont, emitting a twinge of skunk is part of the appeal. US breweries that specialize in saisons, like Jester King in Austin, Texas, have begun packing some of their beers in green glass—going against the industry norm—in the hopes of re-creating the classic style. While met with raised eyebrows at first, their practice is quickly winning over the skeptics.

Glass beer bottles will likely never disappear from shelves, but some breweries are trying out alternatives. A few have released glass-shaped aluminum cans, which they still call bottles. Others are experimenting with specialty materials, for instance the European brewer Carlsberg, which released a bottle-shaped package made from wood pulp that is designed to hold beer for several months without altering the flavor, and is also designed to biodegrade in landfills. Noble as those motivations are, the package is unlikely to take off as a major trend, at least in the immediate future.

I typically think of a bottle (or can) as a delivery device from the brewery before the beer is poured into a glass at home. It's true that now and again I enjoy drinking directly from the bottle. I like the feel of it in my hand, gripping the neck between my thumb and forefinger, and swirling it around while reading at a bar. I like watching other people pull the labels off their beer bottles, out of either nervousness or boredom, or as a challenge to see if it can come off in one piece. I enjoy the solid clink bottles make when bounced together, and the triumphant clatter of an empty being tossed into the recycling bin. That sound does a hangover no favors, but it is an auditory reminder that a good time happened.

Perhaps the thing I like best about the bottle, though, is the cap. Consider for a moment the lowly bottle cap. Before you toss your next bottle cap in the trash can, take a good look at it. In the same way that canned beer has turned the label into a canvas, brewers are using the small space atop the bottle cap to get creative. Once upon a time the caps were a solid color—gold, silver, or black. Today they feature brewery logos, patterned designs, the year a beer was bottled, or even promote a cause or charity. Each September, during National Prostate Cancer Awareness Month, hundreds of brewers add to their bottling runs more than two million blue bottle caps with the logo of Pints for Prostates, a charity that encourages cancer screenings.

Now turn the cap over. When you look at its underside, you might be rewarded with a little saying, picture, or piece of information. Breweries use that circular bit of space to foster interaction with you during your beer-drinking experience. Some of the creations are downright inventive. For decades National Bohemian Beer has printed picture puzzles on their caps that translate into recognizable sayings. They're so popular that websites devoted to solving the riddles have cropped up. In a similar vein Full Sail Brewing has taken to stamping pictures of rocks, paper, or scissors on the underside of their Session Lager beer caps, turning a casual outing with friends into a friendly competition. Magic Hat Brewing and Anderson Valley Brewing have featured witty or insightful sayings under their caps for years, much to the delight of regular customers. Samuel Adams uses the space to tout its many awards. In fact there are now clubs dedicated to collecting the metal rounds.

The top-selling beers in the country—Bud Light, Coors Light, Miller Lite, and Budweiser—all use bottles equipped with screw tops. It's a matter of convenience, a way to get to your beer without the need for pesky tools or a counter top. Most small breweries today use the crown cap because it can form a tighter seal with the glass, keeping oxygen out of the beer, and because it adds a perceived level of sophistication as well.

In some countries you might encounter bottle caps with pull tabs, similar to a can top. They are a fun hybrid between the two but haven't taken off in the United States, in part due to cost (they are more expensive, thanks to the extra pieces) and simply because they are unnecessary.

If the bottled beer you're drinking doesn't have a screw top, you'll need a bottle opener. Bottle openers were born out of necessity. In the early 1890s William Painter of Baltimore received patents on a bottle-sealing cap, the device that keeps liquid and carbonation in a bottle. In 1894 he received a patent for a "capped bottle opener," and the marriage has lasted.

Little has changed over the years in the mechanics of opening a bottle, except when it comes to creativity. Bottle openers come in all shapes, sizes, and materials, and personal taste and expression count for almost as much as functionality. Some drinkers prefer small and practical, for example, a bottle opener that fits on a key ring. Others like wall-mounted openers, or ones designed to be conversation pieces. There is no shortage of openers these days. One version I tend to avoid—and others agree with me—is the shoddy plastic key-ring opener. Typically emblazoned with a brewery logo, they bend more than the caps themselves, are often clunky, and, sadly, usually take more than one try to achieve proper leverage.

Luckily, nearly every household that appreciates good beer is likely to have a few good bottle openers lying around, and some in rather unexpected places. It seems a bottle opener can be incorporated into almost anything. There are flip-flops with openers embedded in the soles, baseball hats with openers inserted into the brims, belt-buckle bottle openers, and even rings with a lifter lip, ensuring you always have an opener on hand. Openers can showcase a person's passion or hobby, depicting everything from Mickey Mouse to the Starship Enterprise or even the pope. One would be hard pressed to find a brewery that doesn't offer an opener with their logo. Materials range from the predictable plastic and stainless steel to the less traditional kangaroo scrotum or bronze from a decommissioned nuclear-weapon system.

Societies and groups exist that promote the collection of bottle openers and bottle caps. Many collectors find unacceptable the natural bend in the cap that occurs when a bottle is opened, so inventors have stepped up and offered a solution. One of my favorites is the Beer Stick, a hefty handle of Appalachian alder with an L-shaped bit of metal at the end that is designed to lift a cap without bending it. It's a durable design, a throwback to when quality was the standard, not the exception.

So what about people who say one bottle opener is just as good as another? An opener that stands out among the rest with a bit of style is something that will start a conversation, just like the beer you've cracked open.

ALTHOUGH THE BOTTLE-OPENER-MANUFACTURING BUSINESS ISN'T in danger of collapse anytime soon, more and more breweries these days offer packages that require no special tool, just a finger. Canned beer is nothing new, of course. It has been around since 1935, when the Gottfried Krueger Brewing Company of New Jersey first packaged their ale in cans. When I started writing about beer back in 2003, very little of the beer-delivery conversation revolved around cans. Sure, the very large breweries were canning their beers, and those were selling quite well, but the can was largely seen as an inferior product designed for low-end beer.

That conversation started to change in the early 2000s, when several smaller breweries, notably Oskar Blues Brewery, began offering their beers in cans and touting its benefits, notably the ability to take cans to places where glassware isn't practical, like the swimming pool, shore, golf course, or campsite. (It's also easier to carry out the empties.) Brewers also talk about how aluminum is lighter than glass. That means there's less weight holding down the trucks that ship beer across towns, states, or the country, reducing the amount of fuel used and exhaust released into the atmosphere.

Most breweries now at least consider using cans, and some even have plans to completely forego bottles and sell their beer only as draft or in the aluminum cylinders. There's a reason why cans are so universally popular: aluminum is one of the most abundant elements on the planet. With great access to the material, it only makes sense that manufacturers will use it. Previously, cans were sold in the tens of thousands at a time, because manufacturers were used to dealing with companies that ordered in those quantities. However, with the growing market of small breweries, for whom those numbers are both impractical and cost prohibitive, can manufacturers such as the Ball Corporation have begun selling smaller lots, making it possible for these newcomers to consider cans.

The rise of smaller breweries has also led to the creation of a complementary niche industry: the mobile canner. Much in the same way that newer breweries lacked the space to store tens of thousands of cans, they usually also lacked space for a permanent canning line. Small breweries were forced to use manual bottling or canning machines that fill four or six packages at a time, a process which is extremely labor intensive and time consuming. Mobile canners are companies with portable, automated canning machines that roll up to a brewery, connect to their bright tanks (a serving vessel at a brewery) or kegs, then wash, fill, and top the cans, getting them ready for the consumer faster. Wild Goose Canning is the Colorado company credited with leading this charge, but today mobile canners exist across the country, and they are in demand and competing with each other for business.

Some concern surrounds mobile canning, and it mostly has to do with maintaining consistency within each beer recipe and packaging run. Larger breweries and ones with fastidious quality-control protocols will run batches of canned (and bottled) beer through all manner of tests, making sure cans are properly filled they've prevented any stray infections, yeast that could continue to ferment beyond the original parameters, or other hazards that could lead to cans bursting. Mobile-canned beer doesn't always go through pasteurization, so it's possible, in the case of a beer style like gose (full of salt) or a New England IPA (full of yeast), that additional fermentation could take place after the beer is canned, putting pressure on the metal and causing it to burst. The same can happen if an infection gets into a beer (or was there all along), which causes additional, unintentional carbonation to build, potentially leading to a rupture. Dozens of breweries in recent years have recalled cans (and bottles too) based on breakage. In most cases fingers are pointed between the mobile canners and the brewers. The ultimate lesson is that at some point, proper care wasn't taken to make sure the packaged beer was ready for prime time. Realizing that this is bad for business, both sides have continued to work toward ensuring that the canned beer that reaches their consumers will be enjoyed without fear of explosion. I can promise you, there are few things worse than waking in the middle of the night to what sounds like a phone book being slammed on a concrete floor in the living room. (That was an exploding bottle of homebrew given to me by a cousin.) Or coming home from an extended summer vacation to find the apartment smelling like a fraternity house on a Monday morning, thanks to a few exploded cans of gose and cider that weren't packaged and sanitized properly from a small professional brewery.

Before we get too much deeper, let's address the other question that comes up most often regarding cans. The answer is no. No, you do not "taste metal" when drinking from food-grade aluminum. It's possible that putting your lips to the rim of a can will give you this subconscious thought, but if it were an actual flavor that could be imparted, brewers (and soda manufacturers) wouldn't use the material. And if you're worried, you can always pour the beer into a glass. Technologies have evolved over the years, and the science behind canning has improved, including making enhancements to the cans' lining, which keeps the liquid from actually touching the metal, preventing corrosion.

The other question I often get asked is if cans are safe for the environment. Yes, generally. They are more likely to be recycled than bottles. However, there is a dark side to aluminum that's not always talked about, and it's called red mud. That's a by-product generated during early stages of mining the material, when it is washed with sodium hydroxide. It is red in color due to the presence of ore, and it is usually stored in secure reservoirs in a sludge form. When fully dried, it can be mixed with soil, or used in concrete or in road construction. Before that point it is toxic, and if it breaks its barriers—as it did in Hungary a decade ago—it can have serious environmental consequences.

There's also the worry of bisphenol A, or BPA, the chemical used in the lining of cans (and also used in other plastics). After studies pointed to the substance as a potential cause of diseases such as cancers and heart troubles, the industry largely moved away from using BPA.

In some ways, cans are the perfect vessel for today's beer industry. Where once the 12-ounce can was the industry standard, now other choices are available. Many brewers opt for 16-ounce, or pint, cans. Others go bigger, at 19.2 ounces or 24 ounces. Cans are even available that can hold more than two full pints (more on that below). The shape of the can generally stays the same across manufacturers, but some improvements have been made over the years, mostly involving the lid. Brewers have created wider spouts to facilitate airflow for pouring or drinking. Then there are cans with fully removable lids, turning the can into a cup—although some states prohibit that packaging (along with tabs that completely pull off) because of litter laws.

Even the way cans are packaged together has changed. Readers of a certain age will remember pro-environmental commercials on television or lessons in an environmental-science class that showed wildlife entangled by the neck in the plastic rings that held cans together. We were instructed to cut them apart before throwing them out. Today, most brewers use snap-on carriers that fit snugly over the tops of the cans. They are recyclable and reusable.

Cans offer two other benefits. They keep sunlight completely away from the liquid, fully preventing the dreaded skunky, lightstruck flavor and aroma. Second, because they are completely sealed, oxygen can't leak in and cause off flavors, or allow carbonation to escape.

With the widespread reintroduction of the can, we're just starting to see how well the packaging holds up to aging beer. The prevailing thought is that because the can has a plastic lining, the beer won't last longer than a year or two. Furthermore, the types of beers being canned—we're still not seeing a lot of barleywines—are designed to be consumed fresh. So for your cellar, stick to bottles for now, at least until more trials are completed.

The practice of canning beers has given local artists a new canvas. With labels that wrap fully around the package, the beer can has become prime real estate for artistic expression. You'll see geometric shapes, cans treated like the pages of a comic book, landscapes, or a story told in words. Employing vibrant colors or stark patterns, the visual of what's on the outside can be just as compelling as the beer within.

Beer cans, especially in recent years, have played a role in disaster recovery. Breweries—especially ones with the capacity for large production runs, such as Budweiser and Oskar Blues—have access to fresh, clean water, and many regularly can drinking water to be delivered across the country after a natural disaster like a hurricane, tornado, or earthquake. They use specially marked cans, which have been a godsend for so many affected.

Personally, I'm a fan of canned beer. I stock up on cans as often as possible. It's not just because of how the can keeps beer fresh, or the cool art on the outside; cans are also a lot easier to carry to the curb on recycling night. Plus, it's good to have so many additional options for your beer-can chicken recipe! (Just make sure your beer is at room temperature before you stuff it into the bird.)

Although you might see cans everywhere these days, and you

know the stigma against them should no longer exist, it's interesting to note that not everyone was quick to adopt the packaging method. Brewery owners Jim Koch of Sam Adams, Deb and Dan Carey of New Glarus, and Sam Calagione of Dogfish Head, for instance, made clear early in their careers that their beers would never see the inside of a can. But you can't stop progress. Seeing the success other breweries were having, hearing the unrelenting pleas from customers (and their distributors), and seeking a way to keep sales climbing, the brewers eventually conceded. These days you can drink Boston Lager, Spotted Cow, and 60 Minute IPA (the flagship beers of the above-mentioned breweries) out of cans. Give the people what they want.

There's a good chance that the first time you visited a brewery you saw a patron come in with a few empty glass jugs in each paw, set the 64-ounce containers on the bar, and watch as they were filled. The customer then walked out with fresh draft beer. Those are growlers, and they are another option for getting beer from a brewery to your house—only involving a little more legwork on your part. In the early days, growlers were small, lidded, tin pails that didn't do much to keep beer fresh for long. You can still spot them in antique stores from time to time. You'll often see the price for growler fills on chalkboard menus, usually just after the price for a pint or a pitcher. It's billed as fresh beer to go, directly from the brewery taps. That sounds amazing, right? Well it is. And it isn't.

If you don't live too far away, and you plan on drinking the beer that day, or within, say, two days max, and you keep the beer properly refrigerated, then it's likely the pints poured from a growler will be close to perfect when consumed. But take any of those steps out of the mix and it's a crapshoot.

Growlers are great in theory. Breweries make a lot of beers that they simply cannot package for takeout in the conventional methods. But sometimes we don't have time to sit down at the bar for a pint, or we want to drink more but we plan to drive home. Or maybe we just want to share a special beer with friends. Growlers can be the solution to all of that. In my opinion, however, there's just too much that can go wrong when it comes to filling growlers, especially if the brewery you're visiting is still using the old-school methods.

What do I mean? Well, that's when a bartender hooks a length of tubing to the tap, sticks the other end into the growler, and opens the line until the beer pours out from the top of the growler. Same as with pouring a pint, the carbonation is released quickly. Then it's trapped inside when the growler cap is screwed on. It's the easiest and most cost-effective way for a brewery—especially a smaller one where fancy equipment might be beyond the budget—to serve you beer to go. It's also not going to do the beer any favors. The oxygen the beer is exposed to, both as it exits the tube and due to any space that might remain at the top of the container, could mean the difference between a flavor that you loved at the bar and one that leaves you scratching your head at home wondering what's changed.

Some breweries, in an effort to combat the disparity, have taken to installing counter-pressure growler fillers. They purge the vessel of oxygen while filling it with beer, similar to how a bottle or can of beer would be filled on the packaging line. The machines are hooked up to the bar's draft system, so they draw from the same kegs as the ones pulling your pint. If I see one of these in use at a brewery, I'll consider getting a growler to go. Otherwise I'll pass.

Crowlers, developed by Oskar Blues Brewery, are basically aluminum growlers (can growlers—get it?) and still relatively new to the beer world. The crowler-filling system is similar to the old tube-on-the-tap system, although in this case the beer is poured directly into a 32-ounce or 64-ounce beer can, much like you'd fill a glass. Then a top is placed on the can, and the machine seams the two pieces together, giving you a big canned beer to go. Thanks to the crowler being filled to the top, there's less space for oxygen to permeate the beer, so it's a good short-term option and also a great alternative to the supposedly reusable glass growlers.

I say "supposedly" because the sad fact is that very few are ever used more than a handful of times. There was a period in my young drinking life when I had a dozen or so growlers from a few breweries scattered across the northeast. In each case, I paid the hefty price for the jug itself before paying for it to be filled with beer. On return visits I'd forget to bring the old ones along and would wind up buying a new growler—repeating that process over and over. Eventually the glass growlers wound up in recycling. I'm seemingly not alone. Brewers across the country have complained that they continually have to order growlers, which seem to be turning into single-use packaging.

Maybe it's a good thing that so many growlers wind up in recycling or as a home bar display. Glass growlers, if not cleaned properly immediately after use, can breed bacteria, mold, and other gunk. Thanks to the thin neck and general size, getting a brush in there for a proper scrubbing isn't the easiest process. Breweries are pretty clear that we, the customers, are responsible for the growlers coming back in good condition. Those that do spot checks won't fill a dirty growler, because it'll harm the beer—meaning you have to purchase a new, clean growler if you want that beer to go. Also, glass growlers are put through the stress of being toted around and knocked into things, and can start to deteriorate or crack after several uses, meaning you'll have to buy a new one anyway. That's why crowler fills are an attractive alternative.

If you think you're going to be the type of beer drinker who regularly gets beers to go, you can invest in a stainless steel growler. These are typically more sanitary (provided you clean thoroughly between each use), more durable, and better insulated than the glass alternative. Some come with fancy bells and whistles, like CO_2 cartridges and tap handles, but simplicity is fine. Many breweries sell stainless growlers branded with their logos, but so long as your state doesn't have a law prohibiting it, you can bring your plain growler in for a fill to take home.

JUST AS A BAR WON'T FILL A DIRTY GROWLER, YOU SHOULD NEVER fill a dirty glass. When you're at home and you have a beer that you're ready to drink, and you've pulled your favorite glass from the cupboard, you want to make sure it's clean. Accomplishing this isn't as hard as you might think. Bars usually use a three-sink method. The first sink is filled with clean, usually running water that has a brush mounted under the water line. The glass is dunked and scrubbed thoroughly, then transferred to the second sink, which is filled with water and sanitizer, for another dunk, and then to a third sink filled with clean water for a rinse. After that it's left to dry.

You're unlikely to want to replicate this system at home, and you don't need to. All you need is access to clean water and a cleaning agent. You can use dish soap. I recommend an unscented one. That green-apple-orchard-scented Palmolive might leave some whiffs behind and impact the beer's aroma. At home and in the office, I use Beer Clean bar glass cleaner, which is a powder, and find that a little bit goes a long way. If you've ever sat at a bar near the sink when glasses are being (properly) washed, you've likely seen it in action. It's also available in the cleaning aisle at most stores. It helps to have a soft-bristle brush as well. After a solid washing, scrub the inside of the glass with the brush while running cool water through it, making sure you wipe away any potential residue. Allow to dry, or use the glass right away. This all sounds obvious, I know, but you would be surprised at how many times I get asked about proper cleaning procedures.

If the glass hasn't been used right away after washing, it's not a bad idea before filling it with beer to give it a splash of water and a swirl to remove any stray particles or dust that may have accumulated. At some bars—and this should be the case at more—you'll see a glass-rinsing station, usually located next to the tap handles, that bartenders use to achieve the same effect. These are pressure-activated water fountains with an attached drain. An upside-down glass is placed on top and pressed down, and water is sprayed inside at multiple angles. Then a quick shake of the wrist removes any large droplets.

We've established that you need the right kind of glass for your beer, and that when you're done pouring there should be room between the top of the foam and the lip of the glass, to help aromas gather and escape. And the glass itself should be clean, at room temperature, and ready to receive beer.

So what's the best way to pour? It's complicated, because it depends on the beer. You've likely seen a draft beer being poured. The bartender holds the glass at an angle under the tap, which is usually hanging down from a wall or tower. When the tap is opened the beer flows down the side of the glass until it reaches the halfway point. Then the bartender gradually straightens the glass to the upright position to avoid spillage, and closes the tap handle when the beer hits the level she or he desires. This is how I poured beer at home from both bottles and cans: a slow, gentle pour down the side, uprighting the glass as the beer neared the top, maybe increasing the velocity a bit to create a head.

For a while I thought I was doing it right, and technically I was, but thanks to Garrett Oliver, the brewmaster of the Brooklyn Brewery, I realized there's a better way. While attending an event at his brewery a few years ago, I watched as he placed a glass on the bar, popped the top of an IPA, and, holding the beer high above the glass, poured it, splashing, directly into the center of the glass. The liquid rose in an agitated manner, yielding a big head of foam and swirling patterns of carbonation, like an angry sea, before settling. He poured about two-thirds of the bottle, saving the rest for later. Oliver explained that the aggressive pour allowed the aromas to bouquet more freely, adding to the olfactory experience. It also demonstrated that beer is tougher than some people might treat it. Now, that's how I typically serve beer at home, unless the beer is bottle conditioned.

Bottle-conditioned beer has had priming sugar added to the bottle before it was capped or corked. The yeast consumes the sugars in the beer during a secondary fermentation, adding new flavors and creating more robust carbonation. This practice is common with many styles, but is generally favored for traditional Belgian styles, like saisons, gueuzes, and lambics. It's a delicate science, because the brewer must know the sugar content of the beer and make sure that each bottle is accurately dosed. Too little yeast and the beer could be flat; too much and the bottle could explode under pressure. You're likely to see newer breweries using the term "bottle conditioned," but they are really adding only a very small amount of yeast or sugar to a beer that has already been force carbonated during the brewing process.

After the yeast is done fermenting, sediment collects at the bottom of the bottle (if stored upright) or along the side (if stored horizontally), in a thick, somewhat sludgy layer. It's delicious but not always visually appealing. There are two ways to handle the sediment. If you're drinking a whole bottle yourself, either use a glass large enough to contain the whole bottle, or plan on pouring two glasses. To pour, start with the gentle, down-the-side technique, until the bottle is two-thirds empty, taking care to avoid disrupting the yeast sediment too much. Then swirl the remaining beer in the bottle to agitate the yeast sediment, incorporating it into the liquid, and do a hard pour into the glass. Yes, the beer will be a little hazy, but that's okay.

Some traditionalists do what they can to keep as much of the residual yeast as possible from making it into the glass. This can mean leaving some liquid at the bottom of bottle, or using a lambic basket. Popular in Belgium, a lambic basket holds a bottle horizontally, and has an opening for the neck, allowing for a relatively smooth pour without disturbing the sediment. Don't expect to encounter one in the States, except in some very select establishments that offer very select beers, or at the home of your serious beer-loving friend.

Speaking of those friends, here's a quick aside: if you're ever at a party where a rare bottle from a brewery like Cantillon, the celebrated Belgian gueuze and lambic brewery, is opened, don't be surprised if there's a homebrewer in the crowd who takes the residual yeast home. It's possible to breathe new life into the microbes, allowing them to ferment yet again. The more revered the brewery, the more likely it is someone will try to homebrew with its leftovers.

Ultimately, when you're out at a bar the bartenders will pour the way they want. Some will do it hard and fast and others slowly and carefully. Either way, the beer will still get to you just fine. At home, if you want to do the slow method, the hard method, or some variation of both, if you can see the beer, smell it, allow it to express itself, and get it into the glass without excessive foaming, you're doing just fine.

HERE'S A SIMPLE QUESTION WITH A COMPLICATED ANSWER: WHAT'S the right temperature at which to serve a beer? If you're at a bar, you generally have to trust the bartender to get it right. But when you're at home, it's much harder to ensure perfection. And as with so many other questions related to this great beverage, it depends on the style. For generations we were conditioned to believe that beer should be served "ice cold," because the dominant American style—the light lager—tastes best when hovering around the freezing mark. As new breweries opened and old beer styles reemerged, however, serving temperature merited a deeper look.

Thirty-eight degrees Fahrenheit. That's the optimal temperature for most beers you're likely to find on draft, and therefore likely to drink. It's also (hopefully) the temperature most bars are set up to deliver. Making sure a keg is stored in a cooler at that temperature for at least twenty-four hours before serving means that it will arrive in your glass at that temperature. It is regarded to be the best temperature for beers such as the American lager. A few degrees colder and the carbonation won't release properly from the liquid, creating a flat sensation. Warmer—even by half a degree—and you're likely to encounter foaming issues because too much carbonation has been released too quickly. You've probably seen this in action: a glass tilted under the tap fills with foam while the bartender either lets it splash out or scoops it out until the glass is filled with beer. In that scenario not only are you getting cheated, but the bars are losing big volumes of liquid that would have otherwise been profit. If you see it happening at a bar, it's a warning sign that something isn't quite right just out of your sight. If it's your home kegerator, it's time to adjust the temperatures, or get out the instruction manual and start troubleshooting.

The American lager starts to impart some less than desirable aromas and flavors as it warms, and hence it should stay cold. But other styles, especially ones with darker malts or floral yeasts, and even some with special ingredients like chocolate or spices, open up and blossom as the temperature increases while you drink. This means that your first sip will taste different from your last one, making the beer more interesting, yes, but also raising the question of why we have to wait for ambient temperature to bring out nuance.

Gabe Gordon of Beachwood Brewing in southern California asked—and answered—that question a few years ago. He, like many others, believed that from the moment we first touch a beer to our taste buds, it should be a sparkling revelation. He built a contraption that he calls the Flux Capacitor (a nod to the time-travel device in *Back to the Future*): a temperature-controlled draft system with elaborate tubing and advanced refrigeration that allows a bar to have multiple taps, all serving at different temperatures. Because while that lager won't display well at 52 degrees, a smoky rauchbier tastes mighty perfect at that temperature. The device also regulates carbonation, and that's important too. Typical draft systems have a one-size-fits-all PSI (pounds per square inch) outflow for carbon dioxide, but that doesn't work for many of the styles popular today, like sours and wild ales. After building the system for his own place, Gordon helped other bars that were serious about good service build and install theirs. It's impressive, like something out of a nuclear lab or a sci-fi movie, and it sparks conversation with customers. The Flux Capacitor is still a rare sight, but beer bars that are passionate about proper service, like ChurchKey in Washington, DC, and Tørst in Brooklyn, are installing the device.

Serving temperature is a little more difficult to dial in at home unless you have a dedicated beer fridge (and if you do, good on you!). Most household fridges are set at a temp of about 40 degrees, and thanks to all the opening and closing you do during the day or when midnight snacking, the actual temperature might fluctuate a bit—meaning the temperature of your beer may change. This isn't the end of the world, because the variations aren't something you're likely to notice when you open a bottle or can and pour the contents into a glass. Bottom line: your regular fridge is a perfectly fine place to store your beer.

But if you want to keep a large selection of beer on hand for long periods of time, you might want to consider a beer cellar: a cool, dark place to store your beer. It's relatively easy to set up a beer cellar, and you don't need a basement. At home I use a bedroom closet carved out of an interior apartment wall. It's insulated well enough and keeps out sunlight. All you need is a similar spot, or a few shelves in your basement or root cellar. Whenever possible, especially with Belgian styles topped with a cage and cork, store the beer on its side, like you would wine. For regular crown-capped bottles, storing upright is fine.

Depending on where you live and the time of year, the natural temperature in your beer cellar can be anywhere from 45 to 55 degrees. This is perfect for long-term storage of certain beer types, like Belgian lambics and gueuzes, and even for higher-alcohol beers, like barleywines, imperial stouts, and old ales. So long as they're not exposed to wild temperature swings, many of these beers, especially if they are bottle conditioned, will age quite nicely for several years. The aforementioned Orval does well for about five years from its bottling point before taking a downward turn, so it's perfect for a cellar. Other beers, such as Harvest Ale, from the Manchester, England, brewer J. W. Lees, continue to evolve and change, becoming more concentrated, complex, and interesting as the years roll into decades. Properly stored in a cellar the Harvest Ale will start as stone-fruit-forward, with plum and fig and even some orange citrus, along with a noticeable alcohol heat from its 11.5 percent ABV. A vintage from 2001 that I drank while working on this book—some sixteen years after it was bottled—revealed a rich, thick toffee verging on a finely fermented soy sauce flavor, with touches of sweet oxidation and cherry pits. It was delightful.

All you need to get a cellar going is the right beer, and I recommend grabbing a few bottles of the same kind for a fun comparison test. Brooklyn Brewery's Black Ops, an imperial stout, is a good example. It's released annually, in a large enough quantity, and is affordable at around twenty dollars per 750-ml bottle. If you can swing it, pick up three bottles of this year's release. Drink one bottle fresh, and take some notes on appearance, aroma, flavor, mouthfeel, and general impression. Tape those notes to the next bottle, and come back to it in a year to repeat the tasting process and see how it's changed. Then wait another year, three years, or even five years to observe its continued evolution. You can do this with any beer, really, but the experiment works best when it's the same beer from a brewery that follows the same recipe each year.

If you get really ambitious, and have the space, you can do vertical flights each year. Bigfoot, the barleywine from Sierra Nevada, is great for this. Line up bottles from this year and previous years, going back annually as far as you can to track how the beer has evolved from its freshest to the one that is longest in the tooth. (The brewery has even begun selling specialty six-packs with four different vintages of the same beer.)

I've had the good fortune of visiting many private beer collections around the country. Some of them contain several thousand bottles, others are more modest. Above all else, hold on to beers that mean something to you personally. I don't advocate hoarding bottles; beers are made to be opened and enjoyed. So if you're going to cellar beers, have a good plan in place for opening them. The worst thing that can happen is that you sit on "special bottles" of beer well past their prime and miss a chance at a great drinking experience.

In this era of endless beer choice, drinking at home is more exciting than ever. A simple trip to a local beer store, or collecting cans and bottles from a brewery road trip, or trading beers with friends through the mail means that when you come home after a long day at work or start that well-deserved weekend break, you'll have a beer to look forward to. Drinking at home means you get to set the parameters. You get the seat you want, the glass you want, and you get to pick the music or what's on the TV.

Some of the best beer nights I've had were at home (mine or someone else's), especially during a bottle share, when everyone brings something different. Maybe they're special bottles that have been held for a while, or a beer from someone's local brewery or region. Or it can be a six-pack of a beer that looked intriguing on the shelf.

There's a relaxed attitude to home drinking, one that lets the beer slip into the background of life: a glass on the countertop as you prepare a family dinner, or on the coffee table while bingeing your new favorite show. Having an enjoyable beer at the ready can turn an ordinary night into something just a little brighter. The best part is that when you're done, you don't have to travel very far for bed.

EIGHT

THE DEATH OF SUBTLETY

I WORRY ABOUT BEER. NOT IN THE SAME WAY I WORRY ABOUT budgeting and paying bills, or the state of government in this country. I worry about beer because although it has been around for thousands of years, think of the immense changes the beverage has gone through globally in the last forty years. It had a long, long infancy and then hit puberty in the 1970s, and today it is an awkward teenager almost ready to become a full-fledged adult. The decisions being made now could affect beer for decades to come.

Beer has always been a beverage of the people, a working-class drink that could also be enjoyed at a fancy wedding. It never got the same respect as wine, but that's changing, thanks to the growing number of breweries, the outspokenness of their fans, and how beer is being presented. A 2014 cover of the *New Yorker* that featured tattooed hipsters swirling and sipping a beer at a bar like it was a bottle of fine wine suggested that different or new beers were only for a certain demographic.

The problem is that the "hipster" demographic is incredibly fickle and actually quite small. "Craft beer," according to the Brewers

Association statistics available when this book was printed, accounts for roughly 12 percent of the overall beer marketplace and about 20 percent in dollar share (the difference between the two numbers is mostly due to the fact that it's more expensive than its mainstream counterparts). That number has remained static for the last few years, meaning that despite each new fad, new trend, or new twist on the IPA, that segment of the beer industry is talking to mostly the same people, and in some ways is losing what made it unique, celebrated, and even respected. I think we need more barleywines, four-ingredient pale ales, stouts without adjunct ingredients, and tripels with traditional yeasts. We need them because they are the foundation of beer, and we need a strong base so that when experimentation happens, it builds upon rather than replaces the basics. I worry about beer because the vocal minority of drinkers—the ones chasing the new, the rare, the local—are the tail that is wagging the dog, and they are sucking some of the everyday joy out of the drinking experience by focusing on the liquid and the Instagram shots rather than on beer in the context of the larger world around them.

I was confronted with this trend a few summers back at the Hill Farmstead Brewery in Greensboro Bend, Vermont. A guy approached me in the gravel parking lot before I had been out of my car for seven seconds. "You getting Damon?" he asked with a sense of urgency. No pleasantries, no greeting. I replied honestly that I didn't know, since I didn't know what that was, and I walked past. I was propositioned by three other gentlemen with the same question before I made it to the front door.

I soon found out that Damon was a bourbon-barrel-aged imperial stout made by the brewery in honor of a beloved dog, and I had shown up on release day. Damon comes in 500-ml bottles, retails for twenty-two dollars, and is limited to one per person, creating the parking-lot black market. On my way out a fifth guy offered me triple the price I'd just paid. Hill Farmstead is a remarkable brewery making many fantastic beers. It brings people to a remote part of Vermont, down long distances of dirt road. Fans and enthusiasts come with high expectations and empty trunks ready to be filled with great beers. I watched as many took selfies with their beer glasses and checked in online, to the great envy of their friends, I'm sure. Here the experience is all about the beer, the bragging rights, and, apparently, the desire to get that second bottle of Damon. I admire Hill Farmstead and enjoy most of the beers they produce. Perhaps my experience would have been different had I not unknowingly arrived on a special bottle-release day.

Many of these hard-won bottles, I suspect, will wind up on beer forums and then packed up and mailed to a new home in exchange for other beers or cash. Trading has never been more popular among beer fans, with message boards—many of them hidden from the public, and accessible by invitation only—teeming with offers of the new, the rare, the local. It's a booming underground business that is sport for some, and infuriating for certain breweries.

Some brewers play into the hype, comfortable in the knowledge that certain beers they sell for forty dollars in their tasting room will go for five or six times that on the internet because of a few opportunistic fans. For other brewers the thought makes their skin crawl. I'm personally against the practice, but I have many friends who participate in the game, and I've been the beneficiary of their kindness if I happened to be around when one of those purchased-above-MSRP bottles was opened.

I'll admit that I got a kick out of the live helicopter coverage on the Chicago news on Black Friday of 2017, when hundreds of people lined up outside not a mall but a liquor store to buy bottles of the annually released Bourbon County Brand Stout from Goose Island. When we see this much hype—something that didn't happen even a decade ago—could some of the phenomena that are more commonplace with wine, like bottles being sold at auction houses, be on the near horizon? The answer is almost certainly yes.

That's ultimately troubling, because although beer is a product and a commodity, it's also a personal experience. Part of the fun is to try new flavors, visit new places, and meet new people, all in the name of beer. Too often, however, there are folks who want to exploit the experience—sometimes by hoarding and then profiting on bottle resales that leave the average drinker out in the cold.

This was apparent at the 2014 Hunahpu's Day, an annual beer festival hosted by Tampa's Cigar City Brewing to celebrate the release of its imperial stout, Hunahpu. In previous years it had been a low-key affair held in the brewery's tasting room and attended mostly by locals. They would stop by, drink a glass, and maybe take home a bottle. But as the brewery's popularity grew and the beer garnered more and more medals and awards, folks soon began making a trip to Tampa for the release. It morphed into a festival: other breweries were invited to come pour their beers, and attendees received bottles of Hunahpu to take home as part of the ticket price. In 2013 more than nine thousand people showed up, causing problems from overflowing toilets to endless lines. Many people left before having a chance to buy the very beer they had come for in the first place. The solution the following year, the brewery decided, was to sell admission tickets online for fifty dollars. While many saw their decision as a reasonable response, others saw it as a chance to exploit customers. When the doors opened that March morning, hundreds of people stood in line with counterfeit tickets. The brewery had planned for a crowd of a certain size, and now it had a much larger one. Rather than sorting it out as the crowd became unrulier, they simply opened the gates and let the chips fall where they may. A lot of folks who paid money for legitimate tickets left without their beer. It was a dark moment in the beer world to witness the deceit and ruthlessness displayed by some. On that day, the "asshole-free" quotient in the industry dropped by a lot.

To its credit, Cigar City, which is now part of Canarchy Craft Brewery Collective, a company controlled by the financial firm Fireman Capital (and that also includes the breweries Oskar Blues, Perrin, and others), has continually tried to improve the experience. It moved the festival off the brewery grounds and into locations that are better equipped for ticketing and crowd control (e.g., Raymond James Stadium, home to the NFL's Tampa Bay Buccaneers). Still, when the gates opened for the 2018 festival, fans—so eager to be first in line for some of the rarest beers being poured—stormed in like it was Black Friday at Walmart, knocking other people down in the process.

I should be clear that I enjoy beer festivals and even some bottle-release days. There's an energy in the air, adults are excited for the event, and generally people are in a good mood. In these cases it's usually less about the intimate beer-sharing experience and more about the social aspect.

Some folks will spend weeks poring over a festival beer list in advance, planning a strategy that will allow them to taste all they desire. Prior to a release day, friends will camp out at a brewery overnight to be among the first to purchase bottles of the new release. These personal interactions strengthen the beer community, allowing folks to make friends with others who share their interests simply by being in the same place at the same time. It's common for beer lovers to discuss their plans with virtual friends as well, sharing thoughts, hopes, tips, and local recommendations online in advance of big events, and leading to a better experience for everyone.

But social media can also take away from the intimate beer experience. This was crystalized for me a few years ago when I visited Cellar 3, the now-closed blending facility run by Green Flash Brewing in San Diego. I was there to catch up with my old friend and local writer Brandon Hernández, but all we could focus on was the couple across the room. The evening was not going well, at least not for the woman sitting at the bar. It was clear that her date was using a mobile app like Untappd, which allows users to "check in" to a beer, rate it in real time, and upload photos and short reviews. It's essentially Twitter with a singular focus for you and your beer buds. It's wildly popular. With every new sip or sample, the man would mention the beer on his phone and then loudly talk about the virtual toasts his choice garnered. His date was eyeing the physical exit.

As breweries have become the new taverns, some of the charms of beer culture are being lost. Conversations still take place in person, but it's often with a beer in one hand and a phone in the other. People hunch over devices rather than soaking in the ambiance and taking time to be present in the moment. So much of our beer experience is personal, but so many are eager to immediately share, brag, or solicit opinions online that the immediate sensations of drinking are lost to pixels, likes, and emojis.

When we slow down—and I include myself in this category—and allow an experience to happen naturally, organically, without worrying about the next social media moment, amazing things can happen. A deeper appreciation can blossom, happy memories can be aroused, new thoughts can develop.

I paid a visit to Cascade Brewing's blending facility near Portland, Oregon, on my thirty-sixth birthday with my companion for the day, Jeff Alworth, a wonderful writer, prolific blogger, and great beer thinker. We were scheduled to take a quick tour, but when head brewer Ron Gansberg greeted us, Jeff smiled and happily muttered, "Oh, this will be fun." Five hours later, when we walked out, our stomachs filled with all manner of aged sours, there was no disagreeing with Alworth's prediction.

Beer needs time in barrels. Yeast can work slowly and deliberately. Fresh fruit used to infuse the beer will mature and reveal depth of flavor. When walking among the thousand-plus barrels in a largely silent space, it's hard not to reflect and slow down. So we did. (It also helped that there was absolutely zero cell-phone service in the warehouse.) Gansberg, with each sample poured from a barrel, talked lovingly of process, aging, and inspiration. He described his career and even opened his bottle cellar, inviting us to taste the brewery's history. He called his coworkers over, and, for several hours on an ordinary Monday afternoon, with each new pop of a cork, he spoke happily, if solemnly, of tradition, his eventual retirement, and the mantle the younger brewers would one day take on.

"In five or six years, the beers we're making now are going to taste this good or better," he said, holding a glass of Cascade Blackberry Ale, poured from the last bottle of the first batch the brewery made in 2007. It was a thrilling moment to witness, and the staff was enthralled. "This, the glorious beginning of why we do this," he told them, "I charge you to carry this on. Make it better . . . and celebrate it always."

His exhortation is important for all of us, and not just when it comes to beer. We need less time in front of screens and more in social settings. We should aim to be like yeast, absorbing all that's around us. We don't always have to hit "refresh" and see what other people are doing. We should savor both the personal and the together moments, the ones we will long remember.

Less than twenty-four hours after leaving Hill Farmstead, I was sitting in an Adirondack chair around a large fire pit outside the Harpoon Brewery in Windsor, Vermont. I was on my second pint of IPA when a fellow patron sat down across from me. "Isn't it a lovely day?" he asked. "How are you?"

It was indeed a gorgeous day. Temps in the low eighties, clear blue sky with only a few high fluffy clouds. A breeze was passing through, and shade was provided by old, big, lush trees. We chatted for a while about jobs and family, about travel and other life experiences. The beer was delicious, but we didn't need to go on about it, because it was only part of the experience. That night a local musician strummed cover songs while people danced. Phones mostly stayed in pockets, and conversation was loud and lively.

I realize that what's important varies greatly for people. For the guys at Hill Farmstead, snagging a second bottle of a special stout was important. For the folks at Harpoon, the ones sitting and chatting, or the ones with six-packs on the lawn playing corn hole, it was about being together and enjoying a stunning afternoon outdoors. The more we focus only on what's to come once a bottle is in hand—the virtual check-ins, the probable profit on reselling a rare release, the bragging for the sake of bragging—the more we depart from the brewer's intentions and the true nature of the beer experience: joy, celebration, discovery.

WE HAVE ONLY A FINITE AMOUNT OF TIME, AND THERE'S A LOT TO enjoy in this world. When it comes to where we consume their beers, brewers are really starting to step up and make sure that customers get the most out of the experience. The breweries that are doing it right—staying true to their mission, making clean, quality beer—are the ones benefiting most these days.

It's part of the economy of time.

I remember my early days of traveling around to breweries. Many, like the Gaslight Brewery, in South Orange, New Jersey, had that British pub feel. Then, as more production breweries opened, they often occupied old (cheap) warehouse space. There was nothing wrong with that, except the tasting rooms felt like an afterthought—there was no attention paid to the fact that some people might actually want to stop in for a visit. I remember visiting one such place where, while in the men's room, I heard my friend's voice saying hello to me. He was looking at me through a hole in the concrete wall. That's an extreme example. Usually, in those early days, patrons might merely be confronted with mismatched chairs, a ratty stereo in the corner, and a sense that a broom hadn't been used in a while. Lighting was bad, the place was too hot or cold, and I had a tough time convincing my then girlfriend (now my wife) to come to a brewery with me.

In the United States we're well past six thousand operating breweries, and thankfully the owners now are putting a lot of thought into how they present themselves. Some are taking pages from winery playbooks and creating spaces where we, the drinkers, can easily spend an afternoon hanging out with friends, drinking either multiple styles or many rounds of the same beer. Kids are playing, dates are happening, and suddenly the customers realize that for the xth Saturday in a row, they've hung out at the local brewery rather than the local bar. Of course, some tavern owners are none too pleased with this development. But as someone who cares about supporting local and drinking fresh, I think it's great that we beer lovers can have a spot that fits our mood.

Brewery owners are thinking about how they present themselves and how they want to be viewed beyond the beer. And while it might be tempting for them to look around and see what competing businesses are doing and how others have been successful, and they might experience a knee-jerk reaction to follow those examples, *most* breweries today are unafraid to forge their own paths.

I was at a brewery in my home state recently, and after a conversation about local zoning and municipal issues, the conversation turned to beer. I complimented the brewer on a hefeweizen brewed with local blueberries. The fresh fruit had turned the beer pink, and despite that, it was popular, the bartender had told me earlier, with even the burliest of guys. It was the closest thing to the Miller Lite those folks were asking for. (Yeah, I was confused too.) It was becoming a best seller. So I had ordered a pint, enjoyed it, paid for it, and paid my compliment to the brewer.

He replied without hesitation that he hated making that particular style. If it were up to him, he'd only make porters and stouts, because that's what he enjoys brewing and drinking.

What was stopping him?

With all the breweries in the United States, there's no law—federal, state, or local—that says a brewer must make a multitude of styles in order to please everyone who comes through the door. No one says a brewery needs to make an IPA. This guy could run the stout brewery of New Jersey. He could make the beers that inspire him and be a go-to source for folks who truly love the style. His sales staff could go after Guinness accounts (the way Left Hand Brewing in Colorado did with their Milk Stout Nitro), spinning the local angle, and he could grow a fine business turning out dark, roasty ales.

I'd argue that we're at a point in beer culture where customers will discover what they like and seek out the places that make it. There's enough of a beer population—for now—to support local breweries that make the flavors drinkers prefer. Because, remember, beer is in a cool spot right now. Especially while there's still a lot of white space that needs to be filled in, brewers should make the beers they want to make. In turn, we'll find the beers we want to drink.

Recently I spoke at a conference in Alberta, Canada, where law changes have led to a boom in brewery openings. I told the folks assembled—the ones who wanted to start their own places—that if they wanted to be known as the saison-only brewery, or as the one specializing in traditional lagers, or be like Black Shirt in Denver that only makes variations on red ales, they could. You don't need to be everything to everyone.

Take the food world, for example. We have bakeries that sell just cupcakes, or just cookies. Dunkin' Donuts largely failed at selling deli sandwiches because their customers wanted only coffee and donuts. If you want tacos, you go to the place that makes good tacos. Likewise, if people want to drink nothing but English milds, and that's what you want to make, be that place, I told the brewers. There's no need to appeal to everyone. They can appeal to the niche inside the niche inside the niche. Earlier in the book I mentioned TRVE in Denver. Known as the "heavy-metal brewery" among the regulars, it's a spot that suits their tastes and is just right. There are countless other examples of breweries that fall into specialized categories of music and culture or geography and history.

When a brewery is true to itself, it will attract customers both local and from far away. I've found perhaps no better example of this than Notch Brewing in Salem, Massachusetts. Of the thirteen hundred-plus breweries I've visited, it's as close to brewery perfection as I've come across. (Sierra Nevada in Mills River, North Carolina, exists in this rarified air as well, but because of its size it's hard to fairly compare the two.) At Notch, longtime New England brewer Chris Lohring spent years finding the perfect spot to make his lagers and pilsners. (He told me that people have turned on their heel and left after asking for an IPA and being told that there are none on offer.) Sparsely but thoughtfully decorated, it sits along a canal, and on days when the sun is shining and your liter mug is full and the pebbled courtyard is humming with conversation punctuated by the call of seabirds, it's hard not to appreciate all the fine details that went into making this a place that will keep you coming back again and again. It might also dawn on you over a few visits that the layout, the decorating, the clean but worn feel of the brick, and the sharp lines on the walls are a reflection of the beers Chris makes.

The same is true of American Solera, a brewery in Tulsa, Oklahoma. Beers that have been lovingly aged in old barrels, foeders, puncheons, and wood are served in a tasting room that is warm with worn-grain accents, chairs, and tables that look comfortable and inviting because they've gotten better with age. There's semidark mood lighting, and low, midtempo music. The place feels like the beer.

Too often—although less and less, thankfully—a brewery's taproom is slapped together at the last minute and has a university frathouse rec-room feel to it. And even though the beer might be great, it's not where I want to spend my time. (I might buy some beer to take home.)

Because here's the interesting thing: a 2016 survey in the United Kingdom found that the British people value time more than they do money. Some 41 percent of respondents to the poll, conducted by the Henley Centre, mentioned time as their most valuable resource, while only 18 percent believed that money was most important. I suspect the same is true in other countries.

It's the economy of time.

Wherever I spend my time drinking beer—be it a festival or a brewery—I want to make sure I'm getting the most fulfillment. A heavy bar tab is likely guaranteed if I walk through the door, but that's still secondary. And I say "likely" because—and this is important—a brewery that is not making quality beer should rethink their priorities. If a brewery has knowingly released a beer that has a major flaw (like diacetyl, oxidation, or DMS)—if they've put it on tap even though it isn't great because dumping the beer would hurt their bottom line—they need to take a hard look at what they are doing. Taking our money in exchange for what they know is a flawed beer is fraud.

Drinkers are savvier than ever. They know when something is off, and with so much choice available, including from breweries right down the street, word will quickly get around. Soon enough customers will be giving the dollars in their pockets to another brewery, and the ones who released a poor beer will have no one to blame but themselves. No matter how flashy the taproom, how on-point the marketing, how golden the promotional tales they've spun, shitty beer is shitty beer. The time it can take to turn around a negative public perception could mean the balance between life and death for a brewery.

A HUGE WEALTH OF POTENTIAL BEER INGREDIENTS IS STILL LOOKING to be discovered, but this richness too often lends itself to gimmicks and fads. Around the time the Celest-jewel-ale (remember, the beer brewed with moon dust?) was released, I found myself at Dogfish's pub in Rehoboth Beach, Delaware. A brewer I know was sitting at the bar looking slightly dejected. When I inquired about his mood, he responded, "We just brewed a beer with an ingredient that didn't come from this planet," he said. "I have no idea what we'll do next."

That's the inherent problem facing all breweries today. Regardless of size, breweries need to keep their existing customers happy while they find ways to attract new drinkers. As we moved away from beer-flavored beer, we went back to the traditional styles, but now we've quickly started chasing trends.

Beer isn't alone in this. Look at vodka. It's a popular spirit, often used as a mixer. Soon enough the distillers realized that they could add cranberry or orange flavoring. Then came vanilla bean, chocolate fudge, cherry pie, and birthday cake. There are now dozens of vodka flavors, and almost all of them tend toward the sweeter end of the spectrum. Variety isn't limited to alcohol either; there are multiple variations on Cheerios, and even Oreo cookies. You don't need to settle for plain old chocolate and cream anymore but can go the orange route, or watermelon, banana split, or, yes, birthday cake. We are simply spoiled for choice in every aisle of the grocery store, including the beer section.

Most people prefer sweet flavors over savory or bitter ones. That's why in recent years we've seen breweries release beers tinged with sweet fruits, including strawberries and kiwi. They've created all-new beer categories to meet this demand, including the New England IPA and the Pastry Stout. New England IPAs are softer, thicker, and more focused on sweeter hop flavors than their predecessors. They are dank and fruity-hop forward, and as clear as a milkshake, with unfiltered hop and yeast particles still floating in the glass when the brew is poured. Pastry stouts are exactly what you think. Stouts in general should already have some chocolate or coffee notes, but now some versions incorporate added sugars and other ingredients that seek to mimic pastry, cake, cookies, or other desserts.

Some older drinkers chafe at these trends, asking what happened to nuance. The drinkers who are just turning twenty-one see it as their generation's statement. Who is right? Who is wrong? Does there even have to be a right or wrong?

The only wrong that I can see is when brewers lose sight of the core mission. When the beverage in your glass no longer tastes or resembles beer, is it fair to call it beer? Has the brewery forgotten its purpose? I think so. Flavored malt beverages, or FMBs, were developed to mimic specific flavors. Popular brands include Not Your Father's Root Beer (and other hard sodas), Twisted Tea, Mike's Hard Lemonade, Four Loko, and the Bud Light "a-Rita" line (think Lime-a-Rita, mango, grape, etc.). Often these are packaged to look similar to beer—in bottles, cans, and on draft—and they appear next to beer on store shelves and taps. But technically they aren't beer. They are wildly popular with consumers, however, so it's only natural that brewers would try to tap into that market.

The brewers who excel in this arena are the ones who add new flavors and layers of nuance without losing respect for what beer is and what it should ultimately taste like: a collection of the four main ingredients. If we lose sight of those qualities, the result will be progress for the sake of progress, and soon enough—I'm talking at least a few generations from now—the beer of the future will look a lot like the theoretical recipes I discussed back in the section on hops.

I DON'T REMEMBER WHEN IT HAPPENED, BUT AT SOME POINT I GAVE up trying to figure out styles of beer. Of course we've learned that beers are descended from lager yeast or ale yeast, but like so much else, it's not always black and white. Kölsch, for example, uses an ale yeast but ferments at lager-fermenting temperatures, and that's traditional, the way it's always been done. Conventionally, the finished product is a brilliant and clear golden yellow, but these days some brewers offer a "black kölsch," an homage to the style that looks more like porter or stout.

The same is true in reverse: there are golden stouts on the market now. The long-established hop-forward IPA, in which the *A* stands for *ale,* now has a cousin in the IPL, with the *L* meaning . . . well, you get it.

A few hundred years ago, breweries throughout Europe employed craftsmen who made traditional beer following exact recipes to strict specifications. A formal law called the Reinheitsgebot regulated beer recipes in Germany, while in England brewers were famously fastidious, making beer to certain specs as a way to avoid excessive taxation in some cases, and generally because their customers demanded a predictable taste. What was once a rigid process, treated like the precision workings of a Swiss watch, has today been left open to interpretation.

There are thousands of breweries operating in this country, and I feel confident estimating that 98 percent of them have an IPA on offer. It's the most popular "craft-beer" style being produced in America today, and it's number one in sales (according to IRI, a market-research firm that tracks industry sales). Enjoying the taste of this hop-forward brew is a badge of honor to some beer drinkers. To tap into the demand, brewers experiment with this style more than any other—but it's a style that lends itself naturally to experimentation. Differences in water chemistry, grain, boil times, hop additions, and varieties of yeast, along with any number of other tweaks a brewer might employ, mean that IPAs are snowflakes. No two are alike.

As a judge a few years ago at the Great American Beer Festival, I had to evaluate the first round of IPAs. Hundreds of entries came in, and our job was to advance only the ones that met the style specifications as laid out in the judges' guidelines, which provided helpful parameters on SRM, ABV, hop content, and problems to look out for, like off flavors. We each received more than a dozen samples—all anonymous—per sitting and had to provide feedback on every one. I was amazed to observe the vast differences between the samples in carbonation, head retention, color, hop aroma, lingering bitterness on the finish, and other qualities.

Beer judging is hard work. Don't let anyone tell you otherwise. It's more than just drinking; it's concentrating on every aspect of a beer, putting it up against predetermined notions, and seeing if it passes muster. Everyone tastes beer in slightly different ways, and in a competition, when it comes time to knock an entry out of the running, pass it along, or award a medal, different judges at the same table tasting the same beer might have wildly different opinions.

As with rap, a West Coast/East Coast rivalry existed in hop-forward ales about twenty-five years ago. It was never violent, thankfully, but it was passionate. In general, West Coast breweries favored including as much lupulin in a recipe as possible, while those on the East Coast preferred malt accents. (Those preferences have since changed.) Following that general rule of thumb, how do you decide that one approach—in a catch-all category like IPA—is better than the other? Ultimately, when winners are announced and you sample a gold-medal champion, it's a lot like pornography: you know it when you see it, or in this case, taste it. Good beer rises above all else, and the IPA style puts hops in the spotlight with other ingredients singing a solid backup.

Aside from awards, though, do we even need styles? Yes and no. For the sake of this part of our conversation let's stick with discussing the IPA family. We in general accept that IPAs are hoppier and boozier than pale ales, but less so than imperial IPAs or double IPAs. Great. We can read the guidelines on each of those styles and be armed with decent information next time we head out for a pint.

The problem is that the brewers themselves don't always follow the specs.

Too often, professional brewers are more than eager to color outside the lines without first seeking to master the original beer. It's a trait that has gotten us to where we are today: history sits on a dusty shelf while the new, shiny things get all the attention. How can we appreciate the latest IPA craze—be it the hazy New England, or the dark-malt-forward black IPA (also known as Cascadian dark ale and American dark ale)—or whatever is coming next if we don't appreciate the beers that came well before, going back to the original Bass Ale from Burton upon Trent, or even earlier?

As a testament to the hardcore nature of IPAs and how much hops they can take, a few years ago a group of San Diego–area brewers collaborated on a beer that they handed out at the 2012 Craft Brewers Conference. It was labeled a "pale ale," but the result was not what one would expect from that style—a balanced ale with around 5 percent ABV—but was rather more like a chewy, dank, smack-in-the-face double IPA. This palate-wrecker is what they consider a pale, one brewer told me. I was terrified to see their interpretation of a double.

Geography certainly plays into tastes. In general, the West Coast, which led the modern IPA charge, is a bit more tolerant for hops than, say, the South, where the modern trend is still coming into bloom. Tastes are generational as well, as we've seen with the recent New England IPA craze among younger drinkers.

Such overwhelming choice is available in beer these days that there's literally something for everyone. A statistic from the Brewers Association's annual report of 2017 claims that the majority of Americans live within ten miles of at least one brewery. If you live in a city like Denver or Portland, Oregon, that number is a lot higher.

An IPA made at one brewery might not suit your tastes, but the same style released by another might hit your sweet spot. I mentioned earlier that, with a few exceptions, I don't care for Belgian-style quads. That's pretty much a sacrilegious thing for a beer lover to say, but it's true. There's something about the general flavor combinations that simply don't appeal to me. When I'm judging, I need to put on blinders and ignore personal preferences and judge the beer on its merits while also leaning on my fellow judges.

Knowing my preferences, I'll study a list for styles when placing an order at a new brewery. I'd rather not spend my money on a drinking experience that I won't like. Still, I'll generally ask for a sample of a beer that isn't among my preferred styles, such as a quad, because maybe my taste buds have changed since the last time I tried it, or maybe the brewer will have done something different that causes me to see the beer in a new way. For the record, Westvleteren XII, generally regarded as one of the finest quads in the world, is something I can happily enjoy, especially when I'm visiting its home country of Belgium.

It comes down to this: sample everything, and then order what suits you best. Be prepared to encounter a huge variety of beers within each style. Just because the chalkboard at the brewery says IPA, that doesn't mean it's going to be a joy (or a disappointment), like one you had previously. If you find an IPA that you like, dissect it using the tasting methods described earlier in this book, and tease out the specifics that suit your fancy—the hop used, the malt content, something else. This will help you make an informed decision next time you order. Personal knowledge is power at the bar.

That said, I realize that with so much choice there's a real inclination to try as many different beers as possible in one night. I'm fortunate to have a career that allows me to sample a wide range of beers; paying a professional visit to a brewery often means jumping between IPAs, barleywines, and stouts in a single sitting. So it took me by surprise when, after a few visits to my new favorite local bar, as soon as I sat down one night, the bartender immediately began pulling a pint of Pilsner Urquell, the Czech beer. He'd observed that even though the bar's rotating list of draft, bottle, and can choices was packed with the latest and greatest, I just wanted the same beer, a simple classic, over and over. I'd reached a renewed appreciation for the subtlety that exists in what most folks would consider a simple beer.

Pilsner Urquell is familiar, and it's consistent. It can blend into the background of the situation. But it can also showcase some nuance when you want to look for it. Spicy Saaz hops bring pepper-like assertiveness to the mug. The pilsner malt can be filling, like country-baked bread. The water is soft and inviting, and there's a slightly fruity yeast ester that weaves between the crispness of a traditional lager.

As drinkers continue to search for the next big thing and brewers try to give it to them, there's still a lot to discover in the old and the familiar. There's a reason why brewers will point to a beer like Sierra Nevada Pale Ale when asked to choose a desert-island beer. It's recognizable, yes, but it has more subtlety and deeper nuance than it's often given credit for. Basic or complicated, Guinness or pumpkin peach pale ale, good beer is good beer. It's the main reason we keep coming back pint after pint.

BREWERIES LOOK FOR WAYS TO REACH NEW CONSUMERS, AND ONCE they get us in the door they try to hold on to us as tightly as they can. That's more difficult these days than it was when the big three brewers ruled the land. Anheuser-Busch could count on Bud drinkers being fiercely loyal and staying that way until the end. Same with Miller and Coors. Powerful brand loyalty isn't uncommon when it comes to consumer products. There are people who prefer Ford trucks over Chevy, Pepsi over Coca-Cola, Chips Ahoy over Nestlé. When few products were available, the beer makers only had to play games of inches. Miller Lite, long the scrappy underdog, came out with their "Tastes Great, Less Filling" ad campaign to woo drinkers away from Budweiser. Budweiser in turn released Bud Light, which became the best-selling beer in America. Coors Light has long been in second place, with the original American Budweiser in fourth, behind Miller Lite. Corona rounds out the top five. However, over the last decade those brands have seen their sales volume fall by numbers that make shareholders nervous. Customers are no longer loyal. Beer drinkers will switch between IPAs and stouts, and drinkers unconcerned with stock prices have no problem drinking a Pabst Blue Ribbon over a Bud. It basically tastes the same, and it's a bit cheaper too!

That's one reason why a few years ago Budweiser released a commercial during the Super Bowl titled "Brewed the Hard Way," which not-so-subtly attacked the upstart brewers that have disrupted its kingdom. It seemed to poke fun at millennials, and at brewers who put fruit in their beer while fussing over the finer details like aroma and flavor. Bud is uncompromising, the ad says. They lager their beer cold, they age it in beech wood, they're a damned American institution and have been since before your great-great-grandparents. You drink us, and you're getting a genuine American product you can be proud of. (Anheuser-Busch InBev is a Belgium-based company.)

Smaller brewers, outraged by the slight, took to social media. They howled at Bud for hitting below the belt, seeming to forget all the shade they've thrown over the years. It was a tempest in a

Teku glass for a while. Then people went back to talking about Spuds MacKenzie, the Bud Light mascot dog.

Somewhat overlooked in all the uproar was the fact that the brewery released the same ad in Spanish and replayed it for months after the Super Bowl on Spanish-language television and radio stations. In addition to Spanish speakers, the brewing giant was also eyeing a new consumer base: immigrants who are chasing the American dream, working hard, making sacrifices, building lives and traditions in their adopted country. This was the beer for them, not that fancy craft stuff, but a product this country helped build, from a company founded by an immigrant that is for immigrants. (AB InBev also aired a commercial featuring a fictionalized immigration story of the brewery's founder.) It was a smart and calculated move by the global brewer, which has also been losing market share in the United States to imported beers, especially brands coming from Mexico.

The bigger the brewery, the more money they will spend to persuade us to give our dollars to them. AB InBev even has an in-house term to measure the success of their advertising campaigns: "share of throat." The first time I heard the phrase I was gobsmacked, but I shouldn't have been. As consumers, we're not friends with that brand. We're not pals or even casual acquaintances. We're the cogs in their moneymaking machine. If one of us doesn't work, they'll look for another to take our place. I know this, but I still found the term insulting, and it's one of the reasons why, when it comes time to open my wallet for beer, I give it to companies who appreciate me for more than just my throat.

It's also one of the reasons why smaller brewers have done so well over the last two decades. They can't afford splashy television ads, so they have to rely on making a personal connection. That generally works. But the customers want something in return, and often it's the three most powerful and dangerous words in beer: new, rare, and local. The latest generation of successful small breweries has trained consumers not to settle for the status quo when it comes to beer. They've taught drinkers to always expect more, something novel. Sales of beers like Sierra Nevada Pale Ale, New Belgium Fat Tire, and Samuel Adams Boston Lager are flat or down, in part because they no longer fit into any of the three buzzword categories.

The companies who do well chasing this trend are usually the smaller ones, the breweries producing a few thousand barrels per year and hyping every release, turning most Saturday mornings into beer-release events. Certain breweries—for example, Bissell Brothers (Portland, Maine), Tree House Brewing (Massachusetts), Trillium (Massachusetts), Other Half (Brooklyn), and Neshaminy Creek (Pennsylvania)—have perfected this marketing strategy. They make lots of money from folks who wait in line outside the brewery for hours before the doors open, as though queuing up to buy the latest phone. Their customers want the status of posting on social media about their early-morning score of a limited-quantity, brewery-only release. The Instagram posts can't happen fast enough.

These beer drinkers want to feel like they are involved with something bigger than themselves or the beer, and they are. A connection is made between brewer and drinker, or between drinker and drinker, and a new kind of brand loyalty is born. Of course, the smart brewers know that they can't be the belle of the ball forever, because this "buzzword" industry model inherently means that someone else will soon become the up-to-the-minute favorite. So while they have a captive audience, they appeal to customers in other ways, perhaps hyping a shared love of music, movies, cartoons, or other pop-culture totems that will prove their street cred with certain groups and ingrain a friendliness that they hope will inspire long-term loyalty.

Looking for that connection is why breweries attempt to rebrand themselves. This is especially true of the Michelob brand, owned by AB InBev. In the 1980s and 1990s it played up its sophistication. It had a distinctive bottle shape with a foil-wrapped neck, and it put out bocks and porters in addition to its traditional lager. When sales flattened and then dipped, Michelob received a rebranding: they created a low-carb, low-calorie lager that marketed itself to health-conscious drinkers, specifically the triathlon set. Michelob Ultra is now synonymous with that activity.

In the end it comes down to giving the consumer what they want. If it's a beer that's been launched into space, great. If it's one for folks who run marathons, you got it. If it's the beer you're convinced is just crazy-bananas good because you stayed in line all night long to get it directly from the source, fantastic. The main thing to remember is that it's just beer. It was never meant to be the all-encompassing giver and taker of life. It is just supposed to be beer.

That fact gets lost for a lot of us, and I include myself in that statement. I've spent afternoons on Twitter and Facebook combating inaccuracies about a beer or brewery, or fighting with someone just for the sake of fighting. When I look around I see others doing the same. Worried about our online reputation, we spend time consulting others' opinions rather than making up our own minds or taking a risk.

It's easy to get caught up in the fever of chasing a new, rare, or local beer without stopping to reconsider and appreciate the classics. This is time spent worrying about what a beer should be or could be rather than what it is, and when that happens we lose sight of what got us excited about beer in the first place. Each new trip to the bar, each new beer opened, is a chance to break that cycle and to focus on the moment at hand.

The reason I enjoy covering the beer industry as well as drinking the product is because of everything else I'm introduced to, in addition to the beer. I learn about raw ingredients and the farmers who grow them, scientists unlocking new discoveries about microorganisms, activists and everyday citizens fighting for clean drinking water. I've been introduced to new music, movies, and TV shows, learned about professions that were unfamiliar to me, and participated in meaningful conversations about life and death and everything in between.

My experience with beer isn't unique because of what I do and who I know. You've likely had similar encounters, and whether you were acutely aware of the beer beside you while they happened isn't the point. The beer was emblematic of the larger spectacle. Each pint we drink today represents a journey that began countless generations ago, one advanced by talent and industry, passion and deep thought. It's the culmination of science, agriculture, creativity, and adventure. Beer, no matter the style, the origin, the maker, or the intent, is certainly a growing part of our cultural and social scene. We've yet to see where this current renaissance will go, if it can be sustained in the long run, or if beer will continue to be beer or will morph into something else. What I do know is that right now is a great time to try new things, find new places, and meet new folks—all while holding a glass of beer in your hand.

EPILOGUE

IT WAS A THURSDAY NIGHT IN JULY IN FARGO, NORTH DAKOTA. I was standing inside a pop-up beer garden where four local breweries were pouring their beer. Vehicle traffic was blocked, and the street bustled with pedestrians thanks to a carnival being held downtown. Businesses had moved some of their wares to the sidewalk, creating an open-market feel. The beer garden occupied an old Goodyear Tire shop that had long since closed, and as pints poured into plastic cups and a guy playing a guitar covered Tom Petty songs, I found myself lost in my thoughts.

How delightfully bizarre this scene is: to see brewers—competing businesses—standing shoulder to shoulder handing out the fruits of their labor. To see local politicians working the room, and folks from every economic class and variety of professions—from public-radio host to rancher, bookstore owner to banker—brought together by locally made beer.

Then I spotted the family. A young wife and husband who looked exactly like people do when they have a one-month-old. Their eyes were both tired and happy, and as the baby slept in his stroller the parents savored their moment of quiet. I watched as each swallow of beer was appreciated with a smile.

The scene spoke to me in a way that was brand new. A week earlier my wife and I had discovered that we were going to be parents for the first time, and my mind raced with thoughts of the baby who would become my daughter. Observing this young family made me feel excited to introduce my child to the larger world, to offer her interesting experiences.

While it can be easy to think of breweries as hulking places filled with industrial equipment (and many are), the grounds, taprooms, and public spaces are largely welcoming. They have long been family-friendly places, going back to the pre-Prohibition days when breweries founded by German immigrants built parks near their facilities (often on the company grounds), where families could relax on weekends, picnic under the shade of trees, and, of course, drink a beer.

I remember being in Milwaukee several years ago at the Lakefront Brewery on a Friday night. They were hosting a fish fry. As the polka band oompahed out the hits, the dance floor filled with three generations of families working off the calories. Couples on first dates sipped beer nervously, and old friends or colleagues gathered for a weekly reunion. Mugs were flowing, and there was a jovial feeling in the air.

I've gotten the truest sense of America by visiting breweries and spending time with the locals. Conversation topics at the bar range from politics to raucous jokes, to stories of loss, love, inspiration, and anger. Some might find my attitude irresponsible, but I'm excited to one day take my daughter to visit these environments. (Of course, we'll also do non-beer activities.)

All this went through my head as I saw the couple and their young child. I realized that beer and beer makers were going to be okay. Beer in America isn't a fad. It's not a blip in our culinary history.

It represents a return to our roots, to patronizing local businesses, to intimate personal experiences, and ones with friends.

THIS ISN'T TO SAY THAT TROUBLE DOESN'T LIE AHEAD. A MONTH before that trip to North Dakota, in 2016, I sat down with Jim Koch from Samuel Adams for the debut episode of my *All About Beer Magazine* podcast, *After Two Beers.* At one point the conversation turned, as it usually does, to the tactics employed by Anheuser-Busch InBev, the largest brewer in the world. The company had been on a buying spree, acquiring several breweries that had once identified as "craft," including Golden Road in Los Angeles, Breckenridge in Colorado, Camden Town in London, and more. All told, today the brewery owns more than a dozen "craft" brands. They've exploited the acquisitions to their advantage, using existing large brewery space to ramp up production at the smaller facilities, and bundling together different styles from their various brands to sell as a package to bars. Julia Herz, of the Brewers Association, has called it the "illusion of choice." Walk into a bar and you'll see an IPA from Goose Island, a stout from Blue Point, an amber ale from Karbach, a Scotch ale from Four Peaks, a lager from 10 Barrel, maybe Stella Artois for international flair, and a handle of Bud Light, because it is still the best-selling beer in America. To the average consumer it looks like quite a variety of options, something for everyone. In reality, all those taps are owned by Anheuser-Busch InBev, and all those profits are going to the one giant corporation. Once AB InBev takes over a bar in this manner, they do everything they can to make sure they never lose those handles to competitors.

Koch told me, in a matter-of-fact tone, "This is the way it is." If Anheuser-Busch InBev wanted to box the top hundred craft breweries in the country out of most bars, they could do it. It wouldn't happen overnight, he said, but they could squeeze most of them out of the industry through competitive pricing, their already deep reach into the marketplace, and a strong grip on the three-tiered distribution system.

Koch's comment stunned me, and I asked him, a smart man who holds a JD and an MBA from Harvard, what could be done to stop it. Nothing, he replied with a shrug. They can do it if they want.

And maybe they will. The company behind Budweiser has never backed down from a fight and has always stepped on the throats of their competition. Throughout the 1970s, 1980s, and early 1990s, AB InBev didn't pay much attention to the smaller breweries because they posed zero threat to their bottom line and seemed like a fad. When they realized that wasn't the case, they went on the defensive, then started playing offense. The result is the huge company you see today: holders of a diversified portfolio with brands that can speak to various demographics without betraying their ownership.

And that's the game of business. You work to be successful, the best, to please your shareholders and keep your employees happy. If Anheuser-Busch InBev wanted to knock out the likes of Bell's or Deschutes, Harpoon or Dogfish Head, New Glarus, Ninkasi, or Allagash, they could. It wouldn't be cheap, and it wouldn't be pretty, but the "craft" segment has stayed at 12 percent of the beer-drinking marketplace for the last few years, and that's not enough to keep the larger players in the segment afloat without growing.

But what Anheuser-Busch InBev and other large brewers can't kill is the neighborhood taproom. The breweries producing a few thousand barrels of beer per year, most of which is sold on the premises or in the local market. The breweries where young families go on a Saturday afternoon or where friends gather on a Thursday night after work. The creative and authentic spaces where camaraderie is encouraged and you get a full artisanal experience, drinking a beer in the presence of the person who made it, in the shadow of the equipment they used.

All this growth in beer happened in a dizzying amount of time. When I was born and was growing up, there simply weren't many breweries in the country where my parents would have felt comfortable letting me visit. In the span of a generation, a whole cottage industry has taken root and sprouted. It's nothing short of impressive.

Yes, there's a lot of change happening in the beer industry. Mergers and sales. Threats to the supply chain, new trends coming out weekly that are dizzying even to the drinkers who like to keep up with these things. But there are more breweries in this country than ever before and a consumer base that continues to grow and be engaged.

We fueled this revolution by simply holding a glass in our hand and asking for something different, something with flavor, something local. Each time we do that we're independent thinkers. Engaged guardians.

There might be tough times ahead, but this is an amazing time to be a beer drinker, and with our support, smart criticism, and enthusiasm for a proud tradition, better days will prevail.

ACKNOWLEDGMENTS

"IS THIS REALLY YOUR JOB?"

That's the question I get asked most often. I consider it a privilege to work as a journalist. It's the only job I've ever had, and each day it gives me the chance to talk with interesting people, learn new things, and visit cool places. Covering the beer industry has been a highlight of an already great career, although sometimes the mornings are difficult.

One of the great pleasures in writing this book was going through old files, calendars, interviews, articles, and memories from my seventeen years covering the beer industry. Putting this book together was like revisiting a giant bar filled with friends, colleagues, brewers, sources, family, and fun folks met along the way.

Since 2004 I've worked at several beer-themed publications, and many of the ideas in this book stemmed from articles and interviews I wrote and conducted over the years. Each assignment helped form my opinion on beer and made me a better drinker, reporter, and consumer. I'm grateful to Tony Forder and Jack Babin of the *Ale Street News* and Tom Dalldorf of *Celebrator Beer News* for giving me an early shot at writing about beer. Nick Kaye, an old colleague from the *New York Times,* became the editor of *Beer Connoisseur Magazine* and brought me on to write and help him out as assistant editor, and I'm thankful for that. At *Craft Beer and Brewing Magazine* I'm fortunate to work with dedicated folks like Jamie Bogner, John Bolton, and Haydn Strauss. I've learned so much about home-brewing culture, and have been able to explore nooks and crannies of the industry that continue to yield interesting finds.

From 2013 to 2017 I served as the editor of *All About Beer Magazine*. It was a great job and shaped my career more than any other position I've held. I will always be eternally thankful to Daniel Bradford, then the owner, for hiring me and sharing his wisdom about the beer industry. Working with folks like Jeff Quinn, Daniel Hartis, Ken Weaver, Adam Harold, Bo McMillan, and Heather Vandenengel made me a stronger editor and taught me the benefit of working in a collaborative atmosphere. Jon Page served as managing editor during the same time, and through him I learned patience and discipline, and the importance of good humor in the face of adversity.

Every Monday at five p.m. I head down to Barcade in Jersey City to record the *Steal This Beer* podcast. My thanks to the staff at the bar, and to my recording colleagues, Augie Carton, Brian Casse, and Justin Kennedy. I've become a better drinker thanks to our sensory analysis and have learned so much from the guests who come by to share their stories and wisdom with us. I've also, strangely, picked up some knowledge on wine.

Every writer should have heroes. When it comes to writing about beer I've been fortunate not only to meet mine, but also to become friends with them. They are sounding boards and trusted confidants. They are also writers, journalists, and authors whose bodies of work have helped shape my own. Any day that I get to hang out with them over beers is a good day: Chris Shepard, Maureen

Ogle, Tom Acitelli, Lauren Buzzeo, Andy Crouch, Jeff Cioletti, and Jeff Alworth.

I'm also fortunate to have met and become friends with talented writers whom I also consider mentors. I encourage you to read everything that Randy Mosher, Ray Daniels, Stan Hieronymus, Pete Brown, and Lew Bryson have put to paper. We are all better-informed drinkers because of them.

As both a writer and an editor in the beer space, I've been fortunate to work with talented and passionate journalists and scribes who toil tirelessly to uncover the truth, tell a good story, and explore new areas of coverage. Thanks to the following individuals for all your support and for encouraging me to work harder: Jason and Todd Alstrom, Tom Bedell, Kate Bernot, Josh Bernstein, Jay Brooks, Jesse Bussard, Thomas Cizauskas, Martyn Cornell, Melissa Cole, Jack Curtin, Christian DeBenedetti, Des de Moor, Gary Dzen, Jeff Evans, John Frank, Oliver Gray, Tim Hampson, Will Hawkes, Danya Henninger, Julia Herz, Brandon Hernández, Adrian Tierney-Jones, Nick Kaye, Ben Keene, Carla Jean Lauter, Jeff Linkous, Jenn Litz, Jimmy Ludwig, Norman Miller, Lisa Morrison, Jon Murray, Josh Noel, Ron Pattinson, Daruss Paul, Roger Protz, Dan Rabin, Evan Rail, Jeff Rice, Erika Rietz, Bryan Roth, Emily Sauter, Nate Schweber, Harry Schuhmacher, Eric Shepard, Joe Stange, Benj Steinman, Bob Townsend, Tom Troncone, Don Tse, Gerard Walen, Tim Webb, Josh Weikert, and Brian Yaeger.

While covering the beer industry I've been fortunate to interview and spend time with dedicated professionals who have opened up their breweries and tapped deep into their knowledge on all manner of subjects to help not only with this book but also with articles, magazines, and podcasts. Thank you to the following: Ali Aasum,

Jodi Andrews, Tyson and Angela Arp, Tomme Arthur, Thomas Peter Barris, Jennifer Baver, Roger Baylor, Larry Bell, Laura Bell, Vilija Bizinkauskas, Chris Black, Erika Bolden, Matt Brynildson, David Buehler, Fred Bueltmann, Sean Burke, Joe Casey, Jimmy Carbone, Sam and Mariah Calagione, Deb Carey, Chris Cohen, Dave Colt, Gwen Conley, Andy Coppock, Peter Crowley, Chris Cuzme, Jeremy Danner, Luc de Raedemaeker, Michele Diamandis, Greg Engert, Jesse Ferguson, Charles and Rose Ann Finkel, Jamie Floyd, Jesse Friedman, Russell Fruits, David Gardell, Paul Gatza, Anne-Fitten Glenn, Matty Hargrove, Craig Hartinger, Chad Henderson, Steve Hindy, Casey Hughes, Mary Izett, Jaime Jurado, Paul and Kim Kavulak, Tom Kehoe, Rick Kempen, Mari and Will Kemper, Jim Koch, Steve Koers, Brian Kulbacki, Ryan Lake, Ashley Leduc, Jeremy Lees, Jeff Levine, Wendy Littlefield, Laura Lodge, Ben Love, Rick Lyke, Bob Mack, Bill Madden, John Mallett, Bill Manley, Garret Marrero, Jace Marti, Tim Matthews, Liz Melby, Sam Merritt, Will Meyers, Derrick Morse, Jonathan Moxey, Rebecca Newman, Scott Newman-Bale, Sean Nordquist, Chuck Noll, Nick Nunns, Jeff O'Neil, Shaun O'Sullivan, David Oldenburg, Garrett Oliver, Jess Paar, Megan Parisi, Steve Parkes, Clay Robinson, Patrick Rue, Ben Savage, Gretchen Schmidhausler, Steve Schmidt, Bryan Simpson, Hugh Sisson, Pete Slosberg, the Soboti family, Mitch Steele, Wolf Sterling, Jon Stern, Jeffrey Stuffings, Terence Sullivan, JC Tetreault, Andy Thomas, John Thompson, Gary Valentine, Brady Walen, Wayne Wambles, Polly Watts, Adam Warrington, Julie Weeks, Byron Wetsch, Jeff Wharton, Ted Whitney, Miles Wilhelm, Matt Van Wyk, Jay Wilson, Jason Yester, and Lisa Zimmer.

Every writer will tell you that you're only as good as your editor. In the case of the book you're holding, Leah Stecher is the critical eye and strong hand behind every sentence, paragraph, and chapter. It's not easy to write a book, but when you have a superb editor like Leah, it makes the whole process just a little more enjoyable. My thanks to everyone else at Basic Books for taking a chance on a book like this with a writer like me. Thank you, Melissa Veronesi. And this book is so much better than the first (and second) draft thanks to the true unsung heroes of journalism, the copy editors. Kelley Blewster is proof that real heroes don't need to wear capes, but are mighty with a red pen. My thanks to her.

I'm also indebted to my literary agent, Jenny Stephens, for making the introduction, and for being a sounding board and generally encouraging person during this process.

Any downtime during the last few years has been spent with friends and family. Yes, there was usually beer involved, but these pleasurable pints were better because of company like Jason Gareis, Katie Zezima and Dave Shaw, Erick and Jaime Lawson, Joe and Rebecca Biland, Bill Malloy, Howie Weber, Pat Battle, Mike Alfonzo, John Kleinchester and Natasha Bahrs, and Os Cruz. And of course my parents, John and Carolyn Holl, my mother-in-law, Teresa Darcy, and my siblings Amanda and Todd Thiede, Bill and Valerie Bronson, Dan Bronson, Tom and Marie Holl, Jill and Beckett Moore, plus the ever-growing list of nieces and nephews.

The best part of my day, every day, is coming home. My wife, April Darcy, is not only a wonderful partner to me and mother to our daughter Hannah (and mutt Pepper) but also a first-class writer and my best editor. Her insight, encouragement, and red pen are on every page of this book (and just about everything else I write), and I wouldn't have even half the career I do without her. Everyone should be as blessed as I am with such a fantastic person in their life.

And finally, to all the readers, podcast listeners, and folks I've met along the way over pints at a bar, at beer festivals and conferences, or at the store: thanks for your recommendations, your tips and advice, and for showing me that what I've chosen to do as a career has such an engaged and enthusiastic audience. When it comes to beer, the best is still yet to come. Cheers!

BIBLIOGRAPHY AND SUGGESTED READING

There is so much to learn about beer, from the nuances of raw ingredients to the detailed process of brewing. I believe that the best way to appreciate beer is to nurture a deep understanding of all aspects of the industry and culture, from beer's history, to proper food pairing, to the economics, to how to brew at home. I highly recommend the following three books to help you forge a deeper relationship with the beer in your glass: *Tasting Beer,* 2nd ed., by Randy Mosher (North Adams, MA: Storey Publishing, 2017); *How to Brew,* 4th ed., by John Palmer (Boulder, CO: Brewers Publications, 2017); and *The Beer Bible,* by Jeff Alworth (New York: Workman, 2015).

In addition, the following books served as resources for this one.

Acitelli, Tom. *The Audacity of Hops: The History of America's Craft Beer Revolution.* Chicago: Chicago Review Press, 2013.

Brown, Pete. *Miracle Brew: Hops, Barley, Water, Yeast and the Nature of Beer.* London: Unbound, 2017.

Carpenter, Dave. *Lager: The Definitive Guide to Tasting and Brewing the World's Most Popular Beer Styles.* Minneapolis, MN: Quarto Publishing, 2017.

Cioletti, Jeff. *Beer FAQ: All That's Left to Know About the World's Most Celebrated Adult Beverage.* Milwaukee, WI: Backbeat Books, 2016.

Cornell, Martyn. *Strange Tales of Ale.* Gloucestershire, UK: Amberley Publishing, 2015.

Daniels, Ray. *Designing Great Beers: The Ultimate Guide to Brewing Classic Beer Styles.* Boulder, CO: Brewers Publications, 1996.

Dawson, Patrick. *Vintage Beer: A Taster's Guide to Brews That Improve over Time.* North Adams, MA: Storey Publishing, 2014.

Hieronymus, Stan. *For the Love of Hops: The Practical Guide to Aroma, Bitterness and the Culture of Hops.* Boulder, CO: Brewers Publications, 2012.

Herz, Julia, and Gwen Conley. *Beer Pairing: The Essential Guide from the Pairing Pros.* Minneapolis, MN: Quarto Publishing, 2015.

Holl, John. *The American Craft Beer Cookbook: 155 Recipes from Your Favorite Brewpubs and Breweries.* North Adams, MA: Storey Publishing, 2013.

Mallett, John. *Malt: A Practical Guide from Field to Brewhouse.* Boulder, CO: Brewers Publications, 2014.

McGovern, Patrick E. *Ancient Brews: Rediscovered and Re-Created Including Homebrew Interpretations and Meal Pairings.* New York: Norton, 2017.

Ogle, Maureen. *Ambitious Brew: The Story of American Beer.* Orlando, FL: Harcourt Books, 2006.

Oliver, Garrett. *The Brewmaster's Table: Discovering the Pleasures of Real Beer with Real Food.* New York: CCC, 2003.

Oliver, Garrett. *The Oxford Companion to Beer.* New York: Oxford University Press, 2012.

Sauter, Em. *Beer Is for Everyone (of Drinking Age).* Long Island City, NY: One Peace Books, 2017.

Steele, Mitch. *IPA: Brewing Techniques, Recipes and the Evolution of India Pale Ale.* Boulder, CO: Brewers Publications, 2012.

Tierney-Jones, Adrian. *Beer in So Many Words: The Best Writing on the Greatest Drink.* London: Safe Haven Books, 2016.

Tierney-Jones, Adrian. *The Seven Moods of Craft Beer: 350 Great Craft Beers from Around the World.* London: 8 Books, 2017.

Tonsmeire, Michael. *American Sour Beers: Innovative Techniques for Mixed Fermentations.* Boulder, CO: Brewers Publications, 2014.

NOTES

CHAPTER 1:

6 *"History is important in the brewing industry"*: Jack McAuliffe, interview with author, May 25, 2010.

8 *"certainly changed my view"*: Michael Lewis, interview with author, June 8, 2010.

12 *In his phone call to me*: Jim Koch, interview with author, September 30, 2011.

CHAPTER 2:

40 *"is the fair-headed stepchild"*: Jamie Jurado, interview with author, June 5, 2011.

52 *"year zero today"*: Stan Hieronymus, email exchange with author, August 14, 2014.

53 *"see a few origin points"*: Carla Jean Lauter, Facebook exchange with author, September 4, 2014.

53 *"need for a bittering agent"*: Gerard Walen, Facebook exchange with author, September 4, 2014.

54 *"MUCH more corn-based"*: Don Tse, Facebook exchange with author, September 4, 2014.

54 *"yellow dye number five"*: Jonathan Moxey, Facebook exchange with author, September 4, 2014.

54 *"be exciting enough to arouse interest"*: Matt Kirkegaard, email exchange with author, September 4, 2014.

59 *"diverse flavors can be had with different yeasts"*: Russ Klisch, interview with author, May 3, 2012.

60 *"wonder if Maier uses a special shampoo"*: John Holl, *AllAbout Beer.com,* November 23, 2015.

CHAPTER 3:

77 *"Generally speaking"*: Patrick Rue, interview with author, November 18, 2014.

78 *"It's not a great, appealing name"*: Michael Tonsmeire, interview with author, November 17, 2014.

CHAPTER 4:

95 *It's all about volume to the customer*: Fergal Murray, interview with author, March 16, 2011.

96 *wants to replace every Stella Artois handle in the greater Denver area*: Ashleigh Carter, interview with author, December 30, 2017.

CHAPTER 5:

127 *A brewery worker in New Hampshire was killed*: US Department of Labor's Occupational Safety press release, October 24, 2012.

132 *"The implication that cask beer"*: Alex Hall, interview with author, April 2014.

139 *"We tend to think of beer as being"*: Randy Mosher, interview with author, June 3, 2011.

CHAPTER 6:

151 *"Demeaning or objectifying women has no place in society or on beer labels"*: John Holl, *All About Beer Magazine,* vol. 38, no. 1.

151 *"chief of the morality police"*: Heat Street, March 2017.

152 *"That might be the worst"*: Luke Gomez, email to author, February 22, 2017.

CHAPTER 7:

163 *"ever-present part of the beer experience"*: Nate Schweber, *All About Beer Magazine*, vol. 36, no. 4.

CHAPTER 8:

193 *"In five or six years"*: Ron Gansberg, interview with author, January 11, 2016.

199 *"We just brewed a beer"*: Author notes, October 4, 2013.

EPILOGUE

215 *"This is the way it is"*: Jim Koch, interview with author, June 16, 2016.

INDEX

Jeff Quinn

JOHN HOLL is senior editor of *Craft Beer and Brewing Magazine* and formerly the editor of *All About Beer Magazine*. An award-winning journalist, he's the author of *The American Craft Beer Cookbook*, has judged beer competitions around the world, and cohosts the podcast *Steal This Beer*. His work has appeared in the *New York Times, Wall Street Journal*, *Washington Post*, *Wine Enthusiast*, and more. Holl lives in Jersey City, NJ.